SPRINGER
LAB MANUALS

Springer-Verlag
Berlin Heidelberg

Ole H. Petersen (Ed.)

Measuring Calcium and Calmodulin Inside and Outside Cells

With 84 Figures and 5 Tables

Springer

C[illegible]
De[illegible]
Th[illegible]
Cr[illegible]
Li[illegible]
UK
e-m[illegible]

[illegible]

Library of Congress Cataloging-in-Publication Data
Measuring calcium and calmodulin inside and outside cells/Ole H. Petersen (ed.) p. cm. – (Springer lab manual) Includes bibliographical references and index. ISBN 3540675361
1. Calcium in the body-Laboratory manuals. 2. Calcium-Analysis-Laboratory manuals. 3. Calmodulin-Laboratory manuals. 4. Cytochemistry-Laboratory manuals. I. Petersen, O. H. (Ole Holger) II. Series.
QP535.C2 M425 2001 572'516 – dc21

http://www.springer.de

Production: PRO EDIT GmbH, 69126 Heidelberg, Germany
Cover design: design & production GmbH, 69121 Heidelberg, Germany
Typesetting: Mitterweger & Partner, 68723 Plankstadt, Germany
Printed on acid free paper SPIN 10669937 31/3130/So 5 4 3 2 1 0

Preface

The study of signal transduction mechanisms has become one of the most important branches of biomedical science, indispensable to understanding the normal actions of hormones, neurotransmitters and other extracellular signalling molecules as well as very many pathophysiological processes including inflammation and cancer.

Cytosolic Ca^{2+} signalling has emerged as a very important branch of the large signal transduction field and has in many ways shown the way forward. This is entirely due to the development of extremely efficient chemical Ca^{2+} indicators and impressive new technologies for detecting the probes.

Two types of intracellular Ca^{2+} indicators have become widely used: probes which change their fluorescence when Ca^{2+} binds and probes that emit light when Ca^{2+} is bound. Fluorescence and luminescence measurements are therefore central to almost all work on cellular Ca^{2+} transport. What has given an enormous boost to Ca^{2+} signalling research is the ability to image the subcellular distribution of Ca^{2+} sensitive fluorescence at high time resolution. This has allowed many investigators to describe in considerable detail the precise spatio-temporal patterns of agonist-elicited cytosolic Ca^{2+} oscillations.

Another highly significant recent development has been the ability to deliver Ca^{2+}-sensitive probes to very specific subcellular compartments and thereby measure the Ca^{2+} concentration inside, for example, the endoplasmic reticulum, the nucleus and the mitochondria. Many important insights into the regulation of cellular Ca^{2+} homestasis have been obtained in this way and there is still scope for enormous further developments.

In this volume, the various techniques for measuring Ca^{2+} in the cytosol, in the various organelles and in the immediate surroundings of individual cells are described in practical details by experts, on the basis of their personal experience. Considering that the protein calmodulin is the main target for Ca^{2+} action in most cell types, a chapter on tracking calmodulin inside cells, using fluorescence technology, has also been included.

Although there are bound to be many important technical developments in the coming years, the principles of Ca^{2+} measurements are now firmly established and it is my hope that this book will continue to be of practical value to all the many scientists who need to consider Ca^{2+} as one of the most important and exciting intracellular messengers.

Liverpool, December 2000 OLE H. PETERSEN

Contents

Part A
Cytosolic Ca^{2+} Concentration Measurements 1

Chapter 1
Principles of Fluorescence Measurements –
Dyes and Hardware Required 3
EMIL C. TOESCU and ALEXEJ VERKHRATSKY

Chapter 2
Simple Principles of Imaging –
Image Files, Image Processing and Image Analysis 47
EMIL C. TOESCU

Chapter 3
Aequorin Measurements of Cytosolic Ca^{2+} 65
C. JANE DIXON and ANNE K. GREEN

Subprotocol 1: Preparing Cells for Microinjection 66
Subprotocol 2: Preparation of Aequorin Stock Solution . . 70
Subprotocol 3: Aequorin Microdialysis 72
Subprotocol 4: Filling Micropipettes for Injection 74
Subprotocol 5: Microinjection of Aequorin Into a Single Cell 75
Subprotocol 6: Signal Detection From a Single Aequorin-Injected Cell 78
Subprotocol 7: Data Presentation 82

Chapter 4
Electrophysiological Measurements
Using Ca^{2+}-Sensitive Microelectrodes 91
ROGER C. THOMAS

Part B
Measuring Ca^{2+} Inside Organelles 103

Chapter 5
Measuring Ca^{2+} in the Nucleoplasm of Intact Cells 105
MARISA BRINI, PAOLO PINTON, PAULO MAGALHAES,
LUISA FILIPPIN, TULLIO POZZAN, and ROSARIO RIZZUTO

Chapter 6
Measuring Ca^{2+} in Isolated Nuclei 131
OLEG GERASIMENKO

Chapter 7
Ca^{2+} Measurement with Luminescent Probes
in the Endoplasmic Reticulum 147
JAVIER ALVAREZ and MAYTE MONTERO

Chapter 8
Ca^{2+} Measurement with Fluorescent Probes
in the Endoplasmic Reticulum 165
ALDEBARAN M. HOFER

Chapter 9
Ca^{2+} Measurements in Mitochondria 187
PAOLO PINTON, ROBERT DRUMMOND,
PAULO MAGALHAES, MARISA BRINI, ANNA CHIESA,
TULLIO POZZAN, and ROSARIO RIZZUTO

Chapter 10
Ca^{2+} Measurements in Secretory Granules 213
OLEG GERASIMENKO

Chapter 11
Ca^{2+} Measurement in Endosomes of Intact Cells 231
JULIA V. GERASIMENKO

Part C
Miscellaneous 249

Chapter 12
Measuring Ca^{2+} Extrusion from Single Cells 251
PAVEL V. BELAN

Chapter 13
Tracking Calmodulin Inside Cells 267
MADELEINE CRASKE

Subprotocol 1: Injection of Labelled Calmodulin into Cells 270
Subprotocol 2: Monitoring Ca^{2+}-Induced Intracellular Calmodulin Movements 275
Subprotocol 3: Monitoring the Ca^{2+}-Calmodulin Reaction in Different Subcellular Compartments .. 283

Subject Index 291

Part A

Cytosolic Ca^{2+} Concentration Measurements

Principles of Fluorescence Measurements – Dyes and Hardware Required

EMIL C. TOESCU and ALEXEJ VERKHRATSKY

Introduction

Fluorescence is by now an almost inescapable every day method for assessing, quantifying or measuring biological processes with a wide array of application in biological sciences. In the context of measurement of free ion concentrations in living cells, and in particular Ca^{2+}, the advances which took place in the last 15 years or so, especially after the publication of the paper detailing the procedures for use of the ratiometric Ca^{2+} dye fura-2 (Grynkiewicz et al., 1985), are truly 'awesome'. As expected, the development of the ratiometric dyes resulted in opening completely new directions of investigation and the general level of interest generated brought, in turn, significant developments in the hardware which supports these type of measurements.

This chapter does not intend to be an exhaustive overview of the fluorescence measurements methodology, but rather aims at pointing at some general principles which should guide the user in deciding which of the various implementations of a 'Ca^{2+} measurement rig' should be better adapted to the intended experimental question raised. The very general scheme shown in Fig. 1 illustrates the basic organisation of the chapter.

Emil C. Toescu, Birmingham University Medical School, Department of Physiology, Birmingham, B15 2TT, UK (*phone* +44-121-414-6927; *fax* +44-121-414-6924; *e-mail* e.c.toescu@bham.ac.uk)
Alexej Verkhratsky, University of Manchester, School of Biological Sciences, 1.124 Stopford Building, Oxford Road, Manchester, M13 9PT, UK

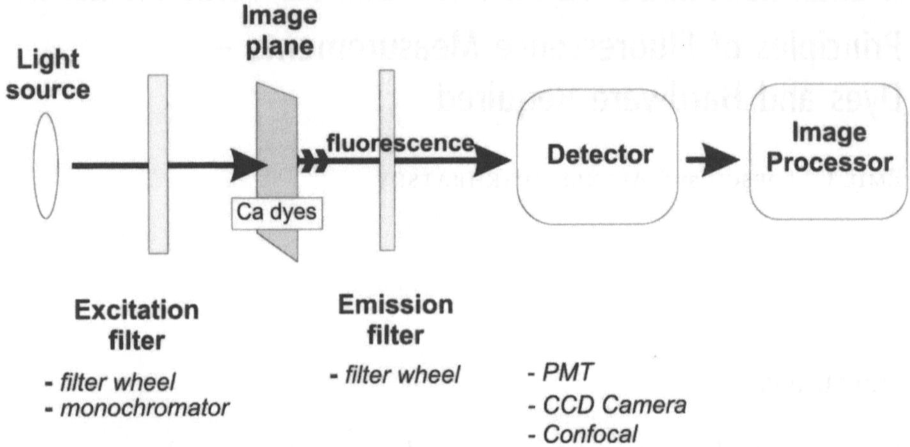

Fig. 1. Basic principles of construction of a Ca^{2+} measurement rig. In the image place, a biological specimen is loaded with a fluorescent dye (see section on **Ca^{2+} dyes**). The dye is excited by light coming from a light source (see section on **Illumination sources**) and the fluorescence emitted is detected (see section on **Detection systems**) and transferred to an image processor unit for storage, display and analysis.

1. Ca^{2+} fluorescent dyes

The quantitative measurements of intracellular calcium concentration reached a routine scale after the introduction of first Ca^{2+}-sensitive fluorescent probe, quin-1 at the beginning of the 80s. Since then, the variety of Ca^{2+}-sensitive probes available for cell biologists has increased dramatically, and a wide range of such dyes are available at the moment (see the complete Molecular Probes catalogue at http://www.probes.com). They all, however, retain the same principle for Ca^{2+} detection: binding of the Ca^{2+} ions to the dye induces changes in fluorescent emission. From a practical point of view these dyes differ in their excitation/emission parameters and their affinity to Ca^{2+} and can be divided into (Table 1):

i. ratiometric and non-ratiometric, and

ii. low and high-affinity.

The choice of the dye is dictated solely by the experimental aims, and the various experimental designs call for a proper Ca^{2+}-probe.

Table 1. Fluorescent calcium indicators

Ca2+ Probe	Spectral properties		Kd	Quantitative[Ca2+]i measurement range
	Excitation	Emission		
Ratiometric/High-affinity Ca2+ indicators				
Fura-2	340/380	510	230 nM	20 - 800 nM
Fura Red	420/480	660	140 nM	20 - 800 nM
Bis-Fura-2	340/380	510	525 nM	100 nM - 2 μM
Indo-1	360	400/500	230 nM	20 - 1000 nM
Ratiometric/Low-affinity Ca2+ indicators				
Mag-Fura-2	340/380	510	25 μM	2 - 100 μM
Mag-Fura-5	340/380	510	28 μM	2 - 100 μM
Mag-Indo-1	360	400/500	35 μM	2 - 100 μM
Non-Ratiometric/High-affinity Ca2+ indicators				
Fluo-3	488	530	390 nM	N/A
Rhod-2	488	570	570 nM	N/A
Calcium-Green-1	488	530	190 nM	N/A
Calcium-Green-2	488	535	550 nM	N/A
Calcium Orange	488	575	185 nM	N/A
Calcium-Crimson	488	615	185 nM	N/A
Oregon Green 488 BAPTA-1	488	520	170 nM	N/A
Oregon Green 488 BAPTA-2	488	520	580 nM	N/A
Non-Ratiometric/Low-affinity Ca2+ indicators				
Calcium-Green-5N	488	530	14 μM	N/A
Calcium Orange-5N	488	580	20 μM	N/A
Oregon Green 488 BAPTA-5N	488	520	20 μM	N/A

1.1. Ratiometric calcium dyes

All ratiometric Ca^{2+} dyes share the same important feature: upon binding of Ca^{2+} ions they change their emission spectra either by shifting the peak excitation wavelength or by shifting the emission spectrum without changes in peak excitation wavelengths. This particular shift allows quantitative measurement of $[Ca^{2+}]_i$ by relating emission spectrum at different Ca^{2+} concentrations. In practical terms, the assessment of the $[Ca^{2+}]_i$ value is performed either by calculating a ratio between the emission values at two-different excitation wavelengths (dual-excitation dyes, such as quin-1 or fura-2), or by rationing the fluorescent signals measured at two different emission wavelength (dual-emission dyes, such as indo-1). The choice between dual-emission and dual-excitation dyes determines the configuration of the experimental set-up, requiring an on-line control over either excitation or emission optical paths (see below). As all these dyes bind Ca^{2+} in a 1:1 ratio, the calibration procedure is straightforward, and involves the measurement of a series of fluorescence ratios (R) at known Ca^{2+} concentrations. The actual value of $[Ca^{2+}]_i$ is determined from the well-known formula (Grynkiewicz et al., 1985),

$$[Ca^{2+}]_i = K_d\beta\ (R - R_{min})/(R_{max} - R);$$

where R_{min} is the fluorescence ratio of Ca^{2+}-free dye and R_{max} is the ratio for dye saturated by Ca^{2+}. The parameter K_d b comprises the value of the affinity of dye to Ca^{2+} ions (K_d - dissociation constant of the dye) and the constant b, determined by the optical properties of the experimental set-up. The K_d, though known for each dye, may vary depending upon the physical properties of the solution (e.g. viscosity, temperature, pH, etc.) and for practical reasons, K_d is usually determined empirically by measuring R for given Ca^{2+} concentration, and using it in the formulae with already known values for R_{min} and R_{max}. With a properly performed calibration, the ratiometric dye technique can provide an accurate estimation of calcium concentration. Conceptually, the K_d value, which represents the half-maximal value of Ca^{2+} dissociation from the Ca^{2+}-dye complex, has important implications. First, as the K_d value stands in the most linear part of the relationship between Ca^{2+} and Ca^{2+}-dye complex, around this value one can obtain the most accurate measurements. Secondly, from theoretical biochemistry, each indicator will have a detectable response in the concentration range of 0.1 x K_d to 10 x K_d. The K_d value could also give an indication of the level of precision at which a particular indicator could provide reliable measurements. For example, $[Ca^{2+}]_i$ recordings using fura-2, with a K_d of 224 nM (at 35-37°C, normal ionic strength, pH= 7.2, 1mM Mg) will be most accurate in the range 150-350 nM Ca^{2+}, and will reliably measure Ca^{2+} in the range 20 nM - 2 μM. Also, it would be realistic to expect that, providing that a thorough calibration procedure has been performed, at around 250 nM, $[Ca^{2+}]_i$ can be measured with a precision of about 5-10 nM, a precision element which will decrease significantly towards the extremes of the measurement range (ie, it would not be realistic to expect that the system could differentiate between $[Ca^{2+}]_i$ values of, say, 975 and 980 nM Ca^{2+}).

1.2. Non-ratiometric dyes

The non-ratiometric dyes do not exhibit any changes in the emission spectra upon binding Ca^{2+}, limiting their reaction to a simple change in fluorescent intensity, when associating with Ca^{2+} ions. The design of this class of dyes (e.g. fluo-3 and 4, Calcium green, Calcium orange, etc.), which are all excited by visible light, was a response to the appearance of the first confocal microscopes which used argon laser with a major line at 488 nm as a light source. The use of visible light for excitation is an obvious advantage, however, as their fluorescence response comprises only a shift in emission intensity, it makes the accurate determination of Ca^{2+} concentration impossible. This is because the intensity of fluorescence is not only a function of $[Ca^{2+}]$ but also depends on an actual dye concentration, and even relatively small variations of the latter, induced by changes of experimental preparation volume or by its movement could greatly affect the estimated $[Ca^{2+}]_i$ values. To circumvent this problem a combination of two dyes, having different emission properties could be used: for instance, cells can be loaded simultaneously with fluo-3 which increases its fluorescence at high $[Ca^{2+}]$ and Fura-Red which decreases fluorescent output while binding Ca^{2+}. Both dyes can be excited at 488 nm and moreover, emission spectrums of Fluo-3 and Fura Red are conveniently separated (emission peaks at 530 nm and 660nm respectively) offering a possibility to segregate signals derived from each dye. Such a technique, even when successfully employed, remains relatively complicated.

1.3. Low and high affinity Ca^{2+} dyes

An important parameter which determines the experimental use of fluorescent Ca^{2+} indicators is their affinity to Ca^{2+}. The latter is determined by dye K_d, which varies between ~200 nM (high-affinity Ca^{2+} indicators, e.g. fura-2, indo-1, fluo-3), 14–20 μM (non-ratiometric low-affinity dyes Calcium-green 5N and Calcium orage-5N) excited by visible light) and 25–35 μM (low-affinity ratiometric UV Ca^{2+} indicators mag-fura-2, mag-fura-5 and mag-indo-1). As discussed, the affinity of Ca^{2+} fluorescent probe determines the actual range of $[Ca^{2+}]$ which can be assessed by the given dye. This certainly dictates the proper use of these dyes: e.g. mag-fura-2 would be the choice when investigating Ca^{2+} dynamics within intracellular stores or hunting for high-amplitude Ca^{2+} transients, whereas fura-2 would be used for characterising calcium changes below 1 mM.

1.4. Cell loading with Ca^{2+} indicators

1.4.1. Preparations for Ca^{2+} measurements

The strategy of practical Ca^{2+} measurements is determined, to a large extent, by the choice of the experimental preparation. Conceptually, four main types of preparations are available for Ca^{2+} recordings (Fig. 2):

- microsomal suspensions;
- cellular suspensions for cuvette measurements;
- freshly isolated or cultured single cells, and
- in situ preparations.

Fig. 2. Types of preparations used for fluorescence measurement of intracellular Ca^{2+}. The preparations used for $[Ca^{2+}]_i$ recordings fall into three groups. The first group, used for populational measurement, performed in cuvettes in microfluorimeters, contains either synaptosomes or cell suspensions, prepared either from freshly isolated or cultured cells. The choice method for loading of cellular/synaptosomal suspensions with calcium probe is the use of membrane-permeable dyes. As the cuvette technique is confined to population recordings, no cellular imaging is possible. The second group is that represented by single cell preparations, which once more could be either freshly isolated or cultured. The cells might be loaded with calcium probe either using a membrane-permeable form of Ca^{2+} dye or intracellular perfusion. The single cell preparations

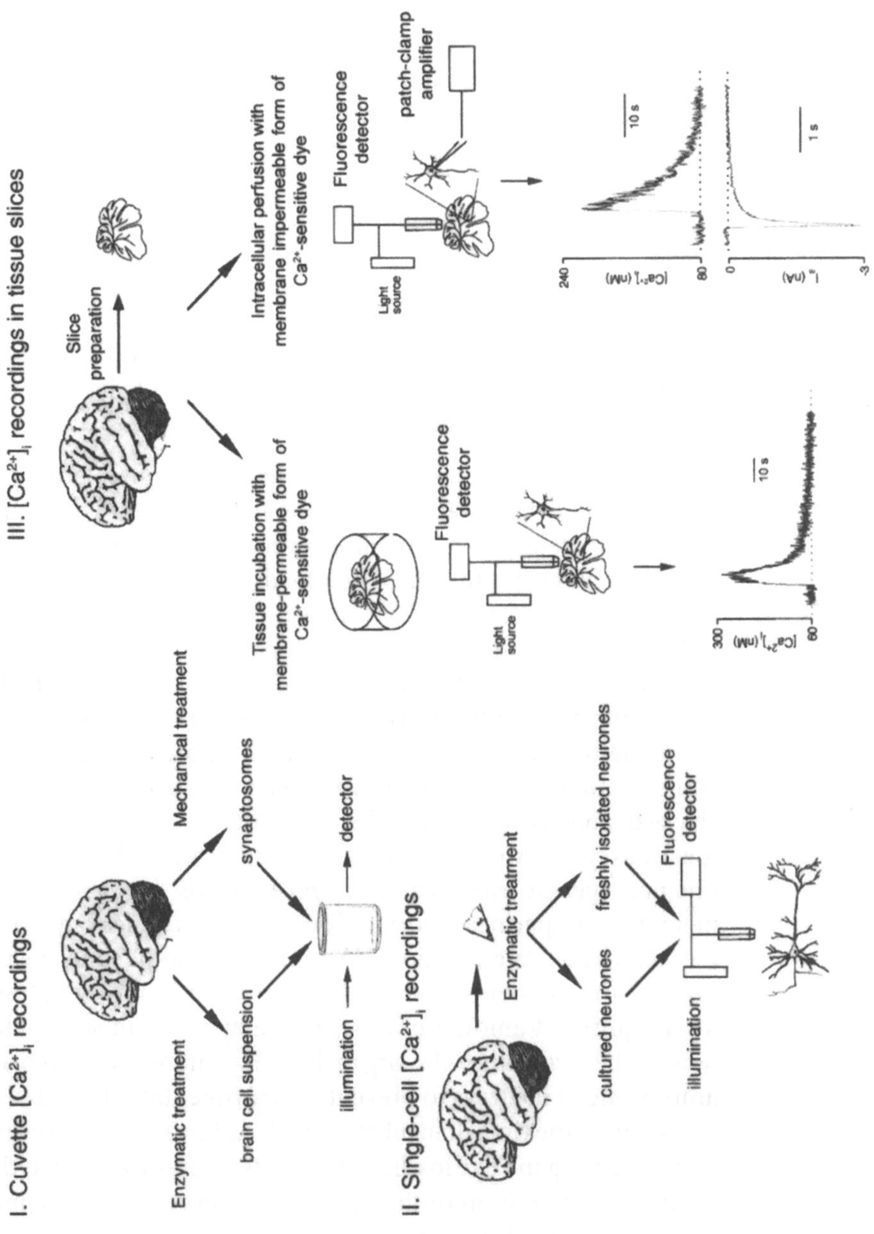

could be used for both, single-detector $[Ca^{2+}]_i$ recordings or $[Ca^{2+}]_i$ imaging. Finally the third group is represented by in situ preparations (e.g. brain slices, muscle strips or tissue fragments). Similarly to previous group cells could be loaded with membrane-permeable dyes or via intracellular perfusion. An obvious advantage is the higher physiological relevance of the results obtained during $[Ca^{2+}]_i$ measurements.

They all have particular advantages and disadvantages, and obviously the choice for the preparation is dictated by the experimental goals. What is important, however, is that different preparations require different methods of introducing calcium indicators into the cytoplasm ("cell loading").

1.4.2. Methods of dye loading

In general, two major approaches for cell loading are used. These are:

1. incubation with membrane-permeable form of the dye, and
2. intracellular perfusion of the cells with artificial solutions supplemented with a membrane impermeant dye (usually in a form of potassium or ammonium salt).

All types of preparations described above are suitable for dye loading with membrane-permeable forms of the dye, and only single cells or in situ preparations offer the possibility for intracellular perfusion. When loading the cells with membrane-permeant form of Ca^{2+} indicator the main obstacle lies in the possible compartmentalisation of the dye within intracellular organelles, which may interfere with recordings. All the membrane-permeant dyes (known as AM form of the indicator) are made by the introduction into the indicator structure of a labile acetoxymethylester group, which renders the molecule charged and hydrophilic. After crossing the cell membrane, the acetoxymethylester (AM) group is hydrolysed by cytoplasmic non-specific esterases ; this process turns the dye into the membrane-impermeant form, preventing thus its leakage from the cell. The de-esterification process requires some time, meaning that the membrane-permeable dye exists in the cytoplasm for a significant amount of time. This may result in accumulation of the dye within the organelles (e.g. endoplasmic reticulum). Some of the ways to prevent compartmentalisation are (1) reduction of the time of incubation with AM form of the dye and (2) performing the dye loading at room temperature. In fact, the compartmentalisation of the dye could be useful, when aiming at measuring Ca^{2+} dynamics in intracellular organelles, and in endoplasmic reticulum in particular (Hofer & Schulze, 1998). In this case, prolonged incubation of cells with AM forms of low-affinity Ca^{2+} indicators results in a loading of ER lumen with the Ca^{2+} probe. Subsequently, the dye can be removed from the cytoplasm by means of intracellular perfusion with a dye-free solution and [Ca^{2+}] changes within the ER can be assessed directly.

Particular care should be applied when the membrane-permeable Ca^{2+} indicators are used for loading the cells in in-situ preparations (e.g. brain slices or muscle segments). From the physiological point of view the in-situ preparation has a clear advantage, as the cells are approached in their natural environment, and results are not significantly affected by the isolation procedure which may often damage the biological object. A number of techniques for loading the insitu preparations with AM form of Ca^{2+} dyes were developed: they are particularly advanced for brain slice preparations (e.g. Kirischuk & Verkhratsky, 1996; Garashuk et al., 1997). The general principle is that the larger the cell of interest, the longer the time needed for loading it with the dye, but the increase in the incubation time also increases the non-specific labelling, thus complicating the experiment.

The second method for dye loading into the cells of interest employs various electrophysiological techniques, when the membrane-impermeant form of the indicator is delivered directly into the cytoplasm either by pressure injection from the microelectrodes or by intracellular dialysis, using whole-cell patch-clamp method. This procedure gives a much better control over the cell of interest and offers simultaneous monitoring of cell electrical activity.

1.4.3. Intracellular concentration of the indicator: Ca^{2+} recordings versus Ca^{2+} clamp

Whichever technique is chosen for cell loading with the dye, the intracellular concentration of the dye is of particular importance. Principally all Ca^{2+}-sensitive fluorescent dyes are Ca^{2+} chelators, and introduction of the dye into the cytoplasm simultaneously adds an exogenous Ca^{2+} buffer, which adds to the endogenous buffer capacity of the cytosol, and thus may contribute to the overall Ca^{2+} dynamics. This consideration underlies two distinct applications of Ca^{2+} indicators: by employing low concentration of the Ca^{2+} probe, one records cytoplasmic Ca^{2+} changes, whereas by loading the cell with high (several μM) concentration of Ca^{2+} dye, one can "clamp" the cytoplasmic $[Ca^{2+}]_i$ and measure the actual transmembrane Ca^{2+} fluxes (Neher, 1994). While the latter technique has a limited application and always requires simultaneous voltage-clamp recordings, the use of the former quite often neglects the possible artefacts associated with Ca^{2+} buffering activity of Ca^{2+} probes. The practical suggestion is simple: while measuring $[Ca^{2+}]_i$, the intracellular dye concentration must be as low as possible, providing it stays within the sensitivity of the detector system.

1.4.4. Calibration and fluorescence background

As mentioned above, Ca^{2+} dye calibration requires measuring the fluorescent intensity at various Ca^{2+} concentrations. The simplest version of the calibration is the in-vitro one, with fluorescence being measured in small droplets of solutions containing the working dye concentration and different levels of Ca^{2+}. A special kit can be obtained from Molecular Probes, containing several precisely calibrated Ca^{2+} solutions. Alternatively, one can use a computer program (e.g., from BioSoft, Cambridge, UK) to calculate the free Ca^{2+} in a solution containing Ca^{2+} and EGTA, for any pH and ionic strength, using well established dissociation constants.

In practice, though, the parameters derived from the in-vitro calibration are not very accurate, since the dye properties in the cytoplasm may vary significantly from those in artificial saline (for this reason some researchers choose to perform the in-vitro calibration in a saline with 2M sucrose, to increase the viscosity of the medium). Thus, the method of choice for physiological experiments is the usage of in-vivo calibration, when the main parameters (R_{min}, R_{max} and $K_d b$) are determined in the cells of interest. For dissociated/cultured cells the calibration, using the calcium ionophores, is widely used. For this purpose loaded cells are incubated, in the presence of Ca^{2+} ionophore (e.g. ionomycin), in media with different Ca^{2+} concentrations. For in situ preparation, the only relevant calibration technique is the intracellular perfusion of the cells of interest with various solutions, specifically designed for determining R_{min}, R_{max} and $K_d b$.

Example

An example of solutions for intracellular calibrations using intracellular dialysis could be as follows: For R_{min} the cells are perfused with (in mM) KCl - 130; EGTA - 10; fura-2 (pentapotassium form) - 0.1; HEPES - 50, pH 7,3 (adjusted with KOH); for R_{max}: KCl - 130; $CaCl_2$ - 10; fura-2 - 0.1, HEPES - 50, pH (KOH) - 7.3; for Kdb KCl - 50; EGTA - 20; $CaCl_2$ - 15; fura-2 - 0.1, pH (KOH)- 7.3; this solution has a calculated Ca^{2+} concentration of 212 nM at 24°C.

Another parameter, which has to be taken into account when working with AM-loaded in situ preparations (e.g. brain slices), is the precise measurement of background fluorescence, which can result from (1) the dye acquired by the tissue which surrounds the cell of interest, (2) the dye compartmentalised into the organelles and (3) the autofluorescence of the tissue or cells. The background fluorescence can be determined either by monitoring the signals from the cell-free regions near the cell of interest (Garashuk et al., 1997), or by intracellular perfusion of the cell with dye-free solution after the end of Ca^{2+} recordings (Kirishuk and Verkhratsky, 1996). The latter procedure can also be useful in experiments on dissociated/cultured cells, when precise calculation of $[Ca^{2+}]_i$ is needed.

2. Hardware for Generating the Fluorescent Signal

2.1. Optical components

For most applications in Ca^{2+} measurement, the image plane is at the stage of the microscope – for work on isolated cells (acutely dispersed or long-term culture) an inverted microscope is preferable, while for work on tissue slices (e.g., brain slices) an upright microscope is required. If simultaneous electrophysiological measurements are required, then the stage space available on an inverted microscope is preferable, again, apart from the work on tissue slices. Fortunately, for work on tissue slices, the new generation of upright microscopes facilitate simultaneous measurements, by providing a fixed stage, separated from the rest of the of the microscope (focusing being done by moving the body of the microscope) and a combination of 'sling in and out' objective turrets with new objectives with tapered ends (e.g., Olympus BXWI50 or Nikon Eclipse600FN) which significantly increase the space available for setting up the patch electrodes. The microscope used needs to be an epifluorescent microscope (a term defining the fact that both the excitation and emission light travel through the objective). The key to the optics in an epifluorescence microscope is the separation of the illumination (excitation) light from the emission fluorescence coming from the excited dye in the sample, and this is achieved through the use of a dichroic mirror

(Fig. 3A), a specially coated mirror onto which the excitation light reflects off into the objective while the emission fluorescence passes towards the eyepiece or the detection system (dichroic means two colours). Each dichroic mirror has a set wavelength value – the transition wavelength value – which is the wavelength of 50% transmission. The mirror will reflect wavelengths below the transition point and transmit wavelengths above this value. However, usually the fluorescent dyes require the selection of a much narrower range of excitation/emission wavelengths, which is achieved by the positioning of adequate, narrow bandpass filters in the excitation and emission light pathways. All these three elements can be incorporated in a single optical object: the filter cubes or blocks (Fig. 3B), which can be matched for specific applications and purchased commercially (two important manufacturers are: Omega Optical (http://www.omegafilters.com) and Chroma Technologies (http://www.chroma.com).

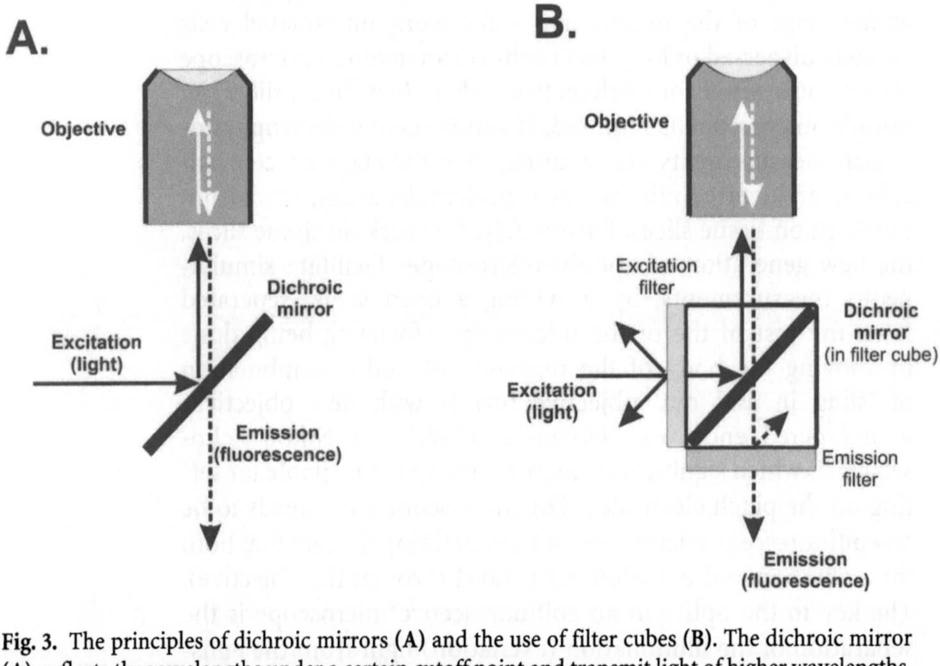

Fig. 3. The principles of dichroic mirrors (**A**) and the use of filter cubes (**B**). The dichroic mirror (**A**), reflects the wavelengths under a certain cutoff point and transmit light of higher wavelengths. In a filter cube (**B**), the dichroic mirror is incorporated with an excitation filter and another emission filter in one single optical object. See text for further details.

Other important terms in an optical filters glossary are:

- **Longpass filters** - filters that transmit all the wavelengths over a specified cut-off point (longer wavelengths). These filters are denoted by the cut-on wavelength at 50% of peak transmission (e.g.- a 505LP filter will have 50% transmission at 505 nm and cut off all the shorter (515 nm) wavelengths.

- **Bandpass filters** - An optical filter that has a well-defined short wave-length cut-on and long wavelength cut-off. Bandpass filters are denoted by their center wavelength and **bandwidth**, which represents typically the separation between the cut-on and cut-off wavelengths at 50% of peak transmission (e.g., a 440/20 BP filter, indicates a central wavelength at 440nm and 50% reduction in transmission for wavelengths 10 nm different on either side of the centre)

Today, many experimental approaches frequently use multiple fluorescent dyes to investigate simultaneously various morphological (e.g., single or multiple immunofluorescence labelling, functional labelling of various cellular compartments, etc.) and/or physiological parameters (Ca^{2+} and pH or Na^{+} measurements, cytosolic Ca^{2+} and mitochondrial Ca^{2+}, etc). This can be achieved either by using several filter cubes (and modern microscopes allow the addition of 3-4 separate filter cubes in their filter turrets) or by use of multiband filter sets, allowing the simultaneous detection of 2-4 combinations of excitation/emission pairs, mainly in the high UV (around 400 nm) and visible spectrum (e.g., FITC/TRITC or DAPI/FITC/TRITC/Cy5).

Before ending the discussion on filters, it is important to note another two types of filters required for microscopic fluorescence measurements. One type is the neutral density filter (ND), the purpose of which is to reduce the amount of excitation incident light, thus protecting both the biological sample and the filter sets. The NDs should be positioned in front of the excitation filter(s), and, depending on the type, will allow light transmission between 0.1% and 90% (sets of different NDs are also available). The second type of filter worth mentioning is the infrared blocking filter (IR), which significantly reduces the infrared energy (600(750)-1200 nm), to be placed either in the excitation path or in front of the detector.

2.2. Illumination sources

In principle, Ca^{2+} measurements using fluorescence microscopy can be run in either one of the two following general arrangements: a) a (quasi) continuous spectrum light source, with excitation interference filters (mounted on a filter wheel) selecting the required wavelength or b) use of a tuneable monochromator to provide directly the required excitation wavelength, thus bypassing the need for specific filters.

In the first arrangement, 2 light sources are commonly used: the mercury and the xenon arc lamps (Fig. 4, top). The mercury lamp has an increased intensity of some of its spectral lines (especially 365nm, but also smaller intensity lines at 313, 334 and 405 nm), which makes it particularly suitable for UV excitation experiments. The spectral lines of the mercury lamp have a width of approximately 10 nm, and in between them the intensity is quite low. In contrast, the xenon lamp has a relatively continuous spectrum in the visible (its major spectral lines are above 800 nm, requiring adequate IR filter protection), but with a relatively low intensity (about 25% of the maximal intensity of its major spectral line at 825 nm) (Fig. 4). The xenon lamp is to be preferred when the spectral characteristics of the dyes and/or specimens are being analysed quantitatively.

Especially for high temporal resolution measurements rapid and precise control of the excitation filters is essential. High-performance excitation filter changers (filter wheels) are difficult to produce. Two main designs are available: driving the filter wheel with a synchronous motor which can be made to rotate at variable speeds, locked to a control frequency (e.g., Cairn Research rotor - which can provide readings at frequencies in excess of 200 Hz). A potential drawback of this system is that it has limited capability of stepping between individual positions and/or maintaining filters in position for different periods of time. The alternative design is to make a digitally controllable stepper motor, which can rapidly move in one direction or another in precisely controlled steps (e.g., the Sutter Lambda filter wheel). The change time (which includes not only the movement but also the stabilisation time in position) could be less than 50 ms per step (i.e., 20 Hz), but is still slow in comparison with the previous arrangement. Overall, for high-speed sampling a constantly rotating filter wheel is preferable, since it

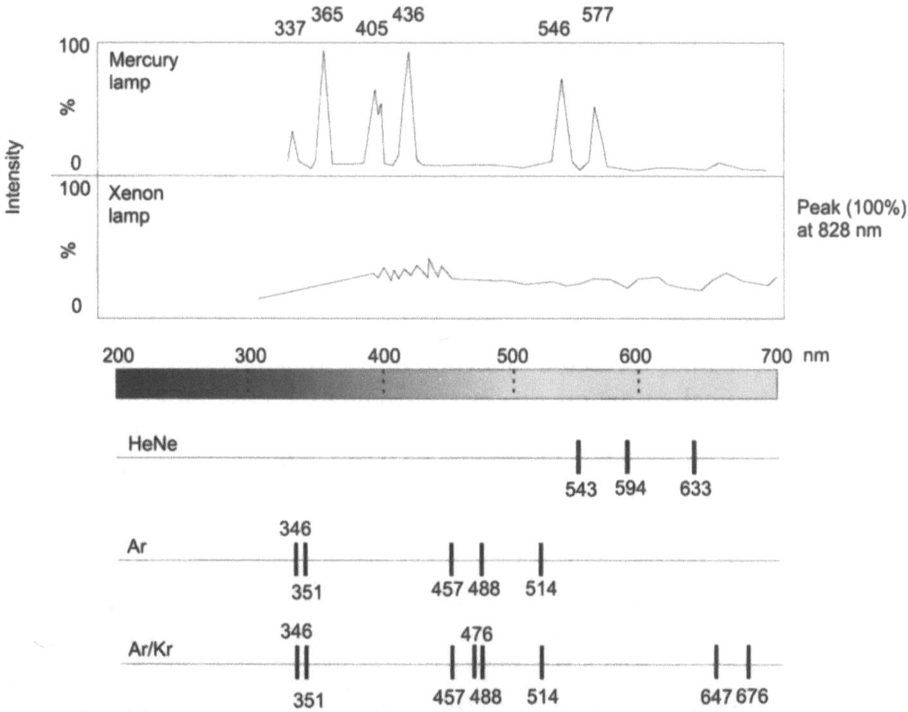

Fig. 4. Illumination sources. Above the diagramatic representation of the spectrum (from UV to visible), the emission spectrum of the two main arc lamp sources are represented, illustrating the maxima of emission and the overall intensity at various other wavelengths. Below the spectrum, the main emission lines of the major laser types used in confocal microscopy are presented.

is much easier to accelerate a wheel to a defined speed and maintain it constantly than to rapidly start the wheel and then bring it to an abrupt stop.

The second arrangement for providing illumination for fluorescent Ca^{2+} microscopy is the use of a monochromator, which offers distinct advantages:

1. free and independent adjustment in wavelength and bandwidth,
2. control of the excitation energy by varying the slit width (bandpass), instead of using ND filters, and
3. the possibility of full spectral scanning of the sample.

Previously, the major problem associated with the use of monochromators was the relative slow time of wavelength change, which led the initial researchers to use dual monochromators, with a chopper mirror which alternately directed the light from each of them, tuned at different wavelengths, towards the sample. The new generation of monochromators uses an acousto-optic device for wavelength tuning, in effect a tuneable diffraction grating (Czerny-Turner configuration). The light from an appropriate source (usually a xenon lamp) is focused on an input slit and is reflected by a concave mirror onto a diffraction grating, which disperses the light beam so that different wavelengths leave the grating at different angles. From the grating the light is focused by a second mirror onto an exit slit forming the final excitation slit. By controlling the position of the grating, only the desired wavelength of the spectrum is focussed on the exit slit. With this arrangement, changes in the wavelength can be reduced to milliseconds, and all the commercially available monochromators (TILL Photonics, Cairn Research, Kinetic Imaging) are capable of this. One special feature separates the Cairn Research monochromator from the rest – and that is the capacity to control, independently and continuously, the width of the entry and exit slits. In taking this approach one always has to balance the bandwidth of the wavelength (roughly, the smaller the slits, the smaller the bandwidth and more precise the wavelength) and the optical efficiency of the transmission, i.e., light intensity (the smaller the slit, the less light gets into and out of the monochromator).

In confocal fluorescence microscopy, as the whole principle of detection (see below) requires light to pass through a single pin-hole which lets through only the light coming from the focus plane of the objective, a much brighter light source is required, hence the need for use of laser light. In biological sciences, three main types of confocal are used: Helium-Neon (HeNe), Argon-Krypton (Ar/Kr) and Argon ion. Each of these light sources will have their specific spectrum lines (Fig. 4). For Ca^{2+} measurements, the more common Argon laser can be used with a variety of visible, single wavelength dyes, whereas the use of ratiometric dyes requires the use of an UV laser source such as the Ar/Kr, with its lines at 346 and 351 nm. The construction of the light input into the confocal setup is highly specialist and is normally implemented through one of the many confocal microscope suppliers (e.g., BioRad, Noran, Nikon, etc.).

2.3. Practical modes of hardware implementation of an imaging setup

1. Multiple excitation/single emission
   This is probably the most common type of Ca^{2+} imaging setup in most laboratories, consisting of a filter wheel or a monochromator as the illumination source, with one filter block in the microscope with a long pass dichroic and a suitable narrow bandwidth emission filter leading to the detector system. This arrangement is of choice when dual-excitation, ratiometric single dyes are to be used (e.g., fura-2). Careful choice of dyes will allow also multi-parameter investigations – for example, fura-2 and BCECF (dye adequate for pH measurements), have different excitation wavelengths (340/380 and 440/490, respectively), but their emission can be collected through the same dichroic/emission (barrier) filter combination (e.g., a LP dichroic 505nm with an emission BP filter 530/30). A very important practical point for this type of arrangement is, taking into account the spectral properties of the illumination source and of the microscope optics, that the visible excitation light will be much brighter than the UV, requiring the addition of ND filters in the BCECF excitation pathway. Another possible protocol, using a similar optical arrangement is to look at the relationship between Ca^{2+} signals and cell viability (using fluorescein or calcein, Ex.494/Em.517) or simultaneous analysis of Ca^{2+} signals and mitochondrial response (using rhodamine 123, Ex.507/Em.529).

2. Single excitation/multiple emission
   This would be the method of choice for the use of the relatively few dual-emission, ratiometric dyes, such as indo-1 (for Ca^{2+} measurement) of JC-1 (for assessment of mitochondrial membrane potential). On the excitation side this arrangement requires just a simple BP filter, but the complication comes in setting up the emission side. This could be achieved either by the use of two dichroic mirrors, arranged in parallel in the emission optical path, each optically coupled to a suitable BP filter, or through the use of a single dichroic mirror and a filter wheel setup in the emission path, containing the respective emission filters. The first arrangement has the advantage that it will allow true simul-

taneous recording of the physiological event measured with minimal registration errors.

3. Multiple excitation/multiple emission
One variant, which we will call 'static', is a result of the developments in optical filter engineering. As mentioned above, today, various filter manufacturers are able to supply multiband filter sets, providing between 2 and 4 simultaneous sets of excitation/emission combinations, mainly used for morphological and immunolocalization studies. The advantage of this method is that it doesn't require any hardware apart from the filter sets (and, if required, a camera) and the image will display simultaneously all the fluorochromes, without any registration errors. The disadvantage, apart from the 'single application' use, is that the brightness could be significantly reduced and that the colour balance is fixed. These facts led to the development of the so-called Pinkel sets, in which the excitation filters are attached to a separate filter wheel. The single BP filters have much better optical properties resulting in brighter images. Also, the fact that each colour is exposed independently allows individual optimization, while the fact that the same dichroic/emission filter is used for all the 2-4 excitation sets minimizes the possible registration shifts associated with changing the whole optical path.
For imaging, another implementation is possible, which gives the most flexible arrangement, allowing true multiparameter physiological recordings. This arrangement, which is currently present in only very few laboratories, has a filter wheel or monochromator as the illumination source and a filter wheel at the emission end. The use of a motorised dichroic turret is possible, but can introduce significant registration and vibration problems; instead the adequate use of a LP dichroic mirror can take full advantage of the separation between the excitation and the emission wavelengths. It is clear that the combinatorial experimental possibilities of such an arrangement are quite large, but could be limited by: a) the number of filter holes available; b) the time taken for each individual image in the cycle sequence; c) the software engineering know-how, to allow

perfect synchronization between excitation exposure time, emission exposure time and camera (or detection) capturing time.

3. Detection Systems

The question of the detection system to be used for a particular application, presuming an instance in which the scientific question comes before that of the technology available, is probably the most important one in setting up a fluorescent Ca^{2+} measurement facility. In general, there are three options available: photomultipliers (PMTs), camera-based imaging and confocal laser systems and a brief overview of their advantages and disadvantages are presented in Table 2.

A detailed analysis of the aims of the experimental project should be carried out before deciding which detection system will be best suited and, especially for camera-based systems which present a large variety of options on the market, which particular system to purchase. Amongst the parameters to take into account, one should address:

- the level of spatial resolution required
- the temporal domain of interest
- the expected dynamic range of intensities in the biological preparation, and
- the types and the number of fluorescent dyes required and their spectral properties.

In the remaining part of this section we will discuss some of the main features and characteristics of each of the main detection systems, insisting mostly on the cameras, where the choice is very wide and requires a better understanding of the technologies involved. The PMT detection systems are becoming more and more obsolete, whereas the discussion on the confocal will be restricted by the fact that, for all its power, there are mainly two configurations, and the possibilities of 'mix-and-match' are very limited.

Table 2. Comparison between the different types of detection systems

Detection system	Advantages	Disadvantages
PMT	cheap to implement	fundamental: absolutely no spatial information
	the highest temporal resolution, approaching resolution of electrophysiology	
	very high detection sensitivity, with low noise	
	allows also spectrofluorimetry (populational, cuvette- or microplate-based measurements)	
Camera-based	very good spatial resolution in the *xy* plane	more expensive
	excellent for single cells (acutely isolated or cultured)	signal contamination from z spatial axis - rendering it less adequate in thicker preparations (although computational (pseudo)confocal solutions are available)
	extreme flexibility in the use of almost any fluorescent dye, allowing a wide range of applications (morphology, immunocytochemical, physiological measurements)	poorer temporal resolution (can be improved at the expense of spatial resolution through binning)
	a wide variety of CCD chips and cameras are available, some built to specific requirements (all this variety can sometimes be a disadvantage in offering too much choice)	
	cameras can be purchased separately and the imaging suites can be built 'in house', with adequate electronic engineering/software development background	
	increased flexibility in matching different cameras to the numerous rame grabber boards and image capture software (NB: not all are dedicated ratiometric capturing programs!)	

Table 2. Continuous

Detection system	Advantages	Disadvantages
Confocal laser	excellent resolution in the xyz plane - method of choice when analysing very small subcellular compartments (e.g., dendritic spines, secretory granules, etc)	very expensive and steep learning curve
	allows choice between image scan and line scan (the latter increasing significantly the temporal resolution)	less flexible in the choice of fluorophores (dependent on the laser source) – ratiometric Ca^{2+} measurements are particularly affected
	allows 3D reconstructions	very strong light leading to potential photo bleaching (dyes) and photo-damage (cells)

3.1. Photomultiplier tube (PMTs)

The classic photomultiplier tube (PMT) was the workhorse in photon detection for many decades and is still widely used today, especially where high sensitivity, ultra-fast measuring rates (temporal resolution) and/or large detection areas are needed. There are two central features of the PMT construction: the photocathode, which generates electrons upon being hit by photons, and the photomultiplication process, due to the existence of a series of 6-9 metallic electrodes (dynodes), held at progressively higher electric potentials, and which are coated with light-sensitive phosphorus. An incident photon strikes the photocathode thereby releasing a photoelectron. This electron is accelerated, in vacuum, to the first dynode by a voltage potential, striking the first dynode with fairly high kinetic energy. This energetic electron will break loose many new electrons in a process known as secondary emission which will each in turn accelerate to the next dynode in the chain. Each dynode collision will increase the total number of electrons by a factor of about 100 resulting in a large pulse of perhaps one million electrons reaching the anode. An important characteristic of the PMT is the **time response** which measures the interval between the arrival of the incident photon and the time of the maximal anode current. Modern PMTs provide overall time response in the range of 20ns, providing output rates of up to 50 MHz.

The spectral response of the PMT is determined on the long wavelength side by the type of photocathode material and on the short wavelength side by the window material (the window being part of the glass encasing in front of the photocathode) and is expressed as quantum efficiency, defined as the number of photoelectrons emitted from the cathode divided by the number of incident photons.

Like all electronic equipment, the PMTs have intrinsic sources of noise, present even in the absence of photons (**anode dark current**). These could be either thermionic emissions emitted by the cathode and/or dynodes, which can be reduced by cooling, ionization of residual gases due to the presence of high velocity electrons and also field emissions which appear when the PMT is operated at voltages near the maximum rated value. For this latter reason it is recommended that the PMTs are operated at voltages around 25% lower than the maximum ratings.

Another important part of the PMT 'package' is the counter device. Modern PMTs have counters able to accumulate well over 10 million counts and which respond to pulses occurring at 10-20 MHz.

A PMT may be operated in either pulse(photon)-counting mode where each pulse of electrons is registered as "an event" or in current mode where the measured current at the output electronics is proportional to the intensity of incident photons. One of the major features of the PMTs is their very high signal-to-noise ratio, which results from the use of **discriminating levels.** This concept refers to the fact that the anode currents under a certain amplitude (corresponding to the various noise sources) and the ones above an upper level are not counted. In this context it is easy to understand that the PMTs are devices in which the incident light is measured not in the amplitude domain but in the frequency domain, i.e., counting events per unit of time. Each 'count' is the consequence of an incident photon and the intensity (amplitude) of the incident light is coded by the number of such counts integrated over the measuring time.

Practical implementations

The combination of extreme light sensitivity (single photon), very short time response and the use of discriminators to reduce to a minimum the effect of the noise defines the main uses of a PMT measuring system - condition of extreme low light and/or the need for very fast (submillisocond) recordings. Obviously these gains in sensitivity and temporal characteristics are at the expense of the spatial information which is lost entirely. For Ca^{2+} measurements, probably the best implementation is the use of user-controlled diaphragms in the light-path leading from the microscope stage to the PMTs, restricting the field of view only to the structures under investigation. This could be done either in the traditional way, using the diaphragm on the optical tube connecting the microscope to the PMT (as in the PTI system) or, by using stage light guides with diaphragms, which will allow 2-3 regions to be monitored simultaneously (as provided by Cairns Research solution).

It is worth mentioning that in addition to the 'traditional' single PMT arrangement, another way to build a PMT setup is to use dual-emission dyes, such as indo-1. This will require one UV light source, but two PMTs with an optical beamsplitter and separate dichroics with BP filters for the two emission wavelengths (405 and 480 nm). Under these conditions, it is crucial here that the two PMTs should be purchased as a pair, with their main characteristics properly matched. The other traditional way is to use.

3.2. Cameras

At present, there are three types of video imaging devices. Two of them, the conventional video tube (e.g., Vidicon) and the image intensifier tube, with the silicon-intensified target (SIT), are obsolete. Currently, almost all the camera-based work in fluorescence imaging uses the charged-couple device (CCD) chips. The array of CCD cameras available on the market could be bewildering and in this section we will try to systematize the various options available and to point to some of the parameters which can help in choosing the proper camera.

The solid-state CCD chip is an array of electrically isolated silicon photodiodes (picture elements, or pixels) of definite dimensions (5-30 μm across), which respond to incident illumination by producing a voltage. The camera's lens focuses the image onto this pixel array. Similar to defining the resolution of conventional photographic film in terms of the size of the grain, the resolution of a CCD chip is measured by the number of pixels in the array and the size of the individual pixels. Although many cameras are available, they will be based only on a limited set of chips (mainly EEV, Kodak and Sony), and the differences in performance between cameras is due to the integration of these chips with the other amplifier and output electronic circuits allowing, for example, manipulations of the sampling intervals, binning, definition of regions of interest (ROI) on the chip, or of the sensitivity, by using either cooling systems or intensifiers (see below).

3.2.1 Classification of CCD cameras

The camera generates a video signal representing the image in front of its lens. For that, the image on the chip needs to be scanned for collecting the information and then this information needs to be read out to the image processor.

A. Scanning modes. In understanding the differences in scanning modes and the recent improvements in this area it is useful to quickly understand the background. In the past, almost all the CCD cameras were based on sensors using the TV scanning formats, mainly because most of the R&D money came from the interest in the development of broadcasting and consumer video markets. This imposed upon the cameras the use of the video standards used in TV. In the 'European' CCIR standard, the scanning speed is 50 fields/sec (imposed by the European use of the 50Hz alternating current) with 625 lines per image. The 'American' standard (RS-170) has a 60 fields/sec rate and 525 lines/image.

Starting from this background, two modes of image scanning have been established.

A1. Interlaced. An interlaced picture consists of two distinct fields of information. On the first pass (field A) the electron beam writes (scans) all odd lines (1, 3, 5, etc.) on the chip. On the next pass (field B), the electron beam scans all even lines (2, 4, 6, etc.) on the panel. The result of this interlaced process is the creation of a complete frame of picture information with every two fields scanned, and thus the frame rate (of full image display) is actually half of the scanning rate. So, if the reported full frame rate of a camera is 1/25 second, the actual field scanning takes 1/50 second. However, the important limitation of this method is that the actual exposure of each field (odd and even) occurs at a different moment in time. Thus, if the camera is capturing a fast changing image, there is the danger that the two fields (A and B) will not register the same image. One solution is to discard the information from one field, but this will come at the cost of reducing the horizontal resolution by half.

A2. Non-interlaced. This method defines a sequential line-by-line scanning process, which results in frame rates of 25 fps. A variant of non-interlaced scanning is called **progressive scanning,** and represents one of the newest technologies. With this method, every line of information on the panel display (1, 2, 3, 4, and so on) is scanned by the electron gun at each pass across the entire panel. The progressive scan cameras support both modes of read out: frame transfer and interline.

B. Read-out modes. This classification refers to the mode in which the CCD chip is shifted from the chip to the output circuit. There are two major principles of transfer:

B1. Transfer cameras – defines the system in which, after the chip has been exposed to an image, each pixel line is read out sequentially. The main drawback of this system is that it is slow, since during the read out the chip needs to be protected from a new incoming image, usually by means of a mechanical shutter, in order to avoid blurring or streaking, especially in the case of fast-changing images. Two implementations of these chips are available: the **full frame transfer camera** and the **frame-transfer camera** - in which only one half of the chip is exposed at any one time, while the other half is being read, under the protection of a shutter. This reduces the area which is imaged by half, but increases the time resolution.

B2 Interline transfer – in which the pixels are arranged in parallel columns on the chip, and the light sensitive columns are separated by blackened out store-regions, used for charge transfer. Thus, the chip's active pixel area and storage register are both contained within the active image area. This differs from "frame transfer" cameras that move all active pixels to a storage register outside of the active area. With this arrangement the electron transfer is very quick, and thus increases the speed of the camera. In some chips, such as the Sony (e.g., Princeton MicroMax) or Kodak, microlenses are placed on the individual pixels, further increasing the light sensitivity. Finally, due to intrinsic properties of the materials used, the interline transfer chips are much more sensitive in the blue/green (400-550) region of the spectrum (where, for example, the emission spectrum for the most common Ca^{2+} sensitive dyes fall - see above).

3.2.2. Important and relevant characteristics of a CCD chip

Here we discuss some of the characteristics of the silicon chips used for light detection in current cameras. However, as will become apparent, the final output features of a camera depend not only on the properties of the detector but also on the various controls and circuits of the camera. Amongst the most important features of a camera are its sensitivity and dynamic range, which are significantly influenced by two important properties of the CCD chip: the quantum efficiency and the pixel size.

Quantum Efficiency (QE) - defines the fraction (percentage) of the incoming photons which are able, at each wavelength, to produce a signal in the ship sensor. Each type of CCD chip has its own unique QE profile. The highest QE is obtained in the so-called 'back illuminated' sensors, which are so designed that the incoming light directly reaches the photosensitive region of the CCD, whereas in the other types ('front illuminated') the light passes through the structures which form the pixels on the CCD, thus reducing the number of photons reaching the detector elements. In relation to the spectral properties, as a general rule, the interline CCD detectors (e.g., the Sony 1300 chip) are more sensitive in the blue-green area of the spectrum (50% QE at ~400 nm in the Sony chip, compared with 5% in a Kodak chip; at ~550 nm both chips have a similar QE (about 40%), whereas at 700nm the Kodak chip still has a QE of 40% (maximal), whereas the QE of the Sony chip is reduced to 15%).

Pixel size defines the actual physical size of the individual sensor element, which can vary between 6 and more than 20 mm. Clearly the larger the pixel size, the better the sensitivity (i.e., increased numbers of photons captured per unit of time), but this is associated with a reduction in the discriminative power of the camera (see Spatial Resolution). The modern cameras allow **binning** (see below), which means that several (multiples of 2) small pixels can be grouped together, for read-out purposes, to form superpixels. Connected with the physical dimension of the pixel is the very important parameter: **well capacity**

of the pixel, which defines the maximal number of electrons which can accumulate in the pixel before spilling over (**blooming**). Small pixels (10 mm or less) have, generally, a capacity of under 10^5 electrons, whereas larger pixels (›20 mm) have capacities of 0.5-0.75 mil electrons. As discussed below, the well capacity is an important factor in establishing the signal to noise features (S/N) and the dynamic range of a camera.

Noise. There are two main sources of noise. One is the **readout noise**, which refers to the random apparition of electrons in the read-out circuit and increases with the readout rate. However, modern chips have very low readout noise (less than 10 electrons at photon-shot noise, which is the result of statistical fluctuations in the photon flux of the incident light. A third source of noise is the **dark current** which is generated spontaneously in the silicon layers of the chip independent of the incident light. The electrons accumulate in the pixels, increase with time and thus might limit the exposure time. An important way of decreasing the effect of the dark current is to lower the CCD temperature (cooled cameras) – modern cameras are run at -10 to -50°C.

Under low-light-level conditions, readout noise exceeds photon-shot noise and the image data is said to be "read-noise limited." The integration time can be increased until photon noise exceeds both read noise and dark noise. At this point the image data is then said to be "photon limited." An alternative method of raising the Signal-to-Noise ratio is to use a technique known as binning.

Dynamic Range (DR). In general terms, the dynamic range describes the number of individual states (gray values), as defined by the level of electron charge in the pixel, that a camera can provide, thus being a measure of the discriminatory power of the camera in respect to the incoming photon flux. The maximal level of the range is given clearly by the well capacity as defined above, whereas the minimal level is given by the readout noise, recorded in the dark. In quantitative terms, the dynamic range is defined by the ratio between these maximum and minimum values.

Thus, for a chip like the Kodak KAF-1400, which has a pixel size of 6.8 mm, a well capacity of 45,000 electrons and a readout noise of 11 electrons (at 1 Mz), the dynamic range is 4091 (45000/11). Formally, the dynamic range is expressed in decibels (dB), using the formula DR=20 x log(well capacity/readout noise), which, for the KAF-1400 chip gives a DR of 72 dB.

The dynamic range assessment also gives an indication of the number of digitization levels that might be appropriate for a given sensor. For the Kodak chip discussed, a 12 bit system, with 4096 levels would be just adequate, whereas for the Sony ICX061 chip (interline chip), which has a well capacity of 18000 electrons with 9 electrons as readout noise (at 1Mz), resulting in a ratio of 2000 (DR=66dB), an 11 bit system (2048 levels) would be sufficient. As a general rule, the cost of a camera increases with increasing dynamic range, so dynamic range requirements should be considered very carefully when selecting a camera. Note further that in an uncooled video camera the read noise is much higher, and the true dynamic range is usually 8 bits or lower.

It is clear from this discussion about pixel size, dynamic range, and well capacity that, in choosing and using a camera, there is an important trade-off between spatial resolution, which is improved by a having a smaller pixel, and dynamic range, which requires larger well capacities, and thus pixels.

Signal to Noise Ratio (S/N). Whereas DR defines mainly the properties of the chip, the S/N concept takes into account the incoming image and reflects the relationship between the signal generated by this incoming image (photon flux) and the noise associated with the capture of the image. The total noise (N_T) that contaminates a captured signal comes from three sources, as discussed above: readout, dark current and the photon shot noise.

It is in the nature of things that each signal carries an inherent noise, due to statistical fluctuations, distributed according to the Poisson distribution, and the relationship between signal and its noise is defined, in these conditions by a square root formula:

$$\textbf{Noise} = (\textbf{Signal})^{1/2} \qquad \text{(i)}$$

Implicit in this formula is also the fact that each type of noise present in the system will result in a signal, which will contaminate the image signal (S_{im}). Thus, the S/N is, in fact, the ratio between S_{im} and the total noise N_T.

Let us analyse in more detail the actual measurements involved in calculating the S/N. What is S_{im}? Each detector element (pixel) receives a flux of photons (the 'image'), measured in photons/sec (F), which is transformed in a flux of electrons according to the quantum efficiency (QE) for the wavelength of the incoming photons. Also, the flux of outflowing electrons depends on the time of exposure (integration time). The electronic signal is digitized, using a specified gain (see below), but we will consider here only the simpler case of a gain of 1. Thus,

$$S_{im} = \mathbf{F} * \mathbf{QE} * \mathbf{t} \qquad \text{(ii)}$$

How do we determine the total noise in the system (N_T)? The simplest way is to use equation (i), in which case N_T is the square root of the total signal (S_T).

$$N_T = (S_T)^{1/2} = (\mathbf{S_{im}} + \mathbf{S_d} + \mathbf{S_{ro}})^{1/2} = (\mathbf{S_{im}} + \mathbf{N_d^2} * \mathbf{t} + \mathbf{N_{ro}^2})^{1/2} \qquad \text{(iii)}$$

where, $\mathbf{N_d}$ is the dark noise current (electrons/sec/pixel), generating the dark noise signal ($\mathbf{S_d}$), which is larger as the time of integration (**t**) increases, and $\mathbf{N_{ro}}$ (electrons/pixel) is the readout noise, determined only by the readout speed (see above). Thus, combining (ii) and (iii), we can obtain a mathematical expression of the S/N value:

$$S/N = (\mathbf{F} * \mathbf{QE} * \mathbf{t})/((\mathbf{F} * \mathbf{QE} * \mathbf{t}) + \mathbf{N_d^2} * \mathbf{t} + \mathbf{N_{ro}^2})^{1/2} \qquad \text{(iv)}$$

As we will see below, the trouble in understanding the derivation of the S/N formula will pay dividends when the methods of improving images acquisition are assessed.

Spatial resolution. Understanding the spatial discriminatory power of a camera, requires the correlation of two concepts. The first concerns the optical limitations imposed by light diffraction on the resolving power of an optical system. Any microscope objective (i.e., the image source for the camera) is characterised by its numerical aperture (NA), defined as the product of the refractive index (r_i) of the medium interposed between lens and specimen (1 for air, up to 1.5 for immersion oil) and the sin value of half the angle of the lens aperture (Fig. 5).

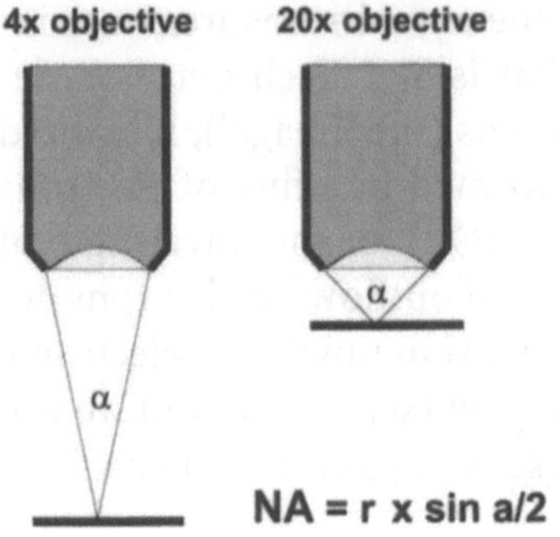

Fig. 5. Principles of measurement of numerical aperture (NA) of a microscope objective. See text for details.

The spatial resolution of an objective, i.e., the smallest resolvable distance (**d**), is dependent on the NA of the objective and the wavelength of the light (**l**), according to the Rayleigh expression:

$$\mathbf{d = 0.61 * \lambda/NA} \qquad \text{(iv)}$$

which indicates that the resolution of the optical system is increased (ie, **d** is decreased) with an increase in the NA of the objective or towards the UV/blue end of the spectrum. As a note, the NA of the optical system should be considered, for a microscope lens, the average numerical aperture for the lens and the condenser. Thus, if the condenser is significantly underfilling the objective with light to increase image contrast, the spatial resolution is decreased.

The second concept which needs to be discussed is that of matching the CCD detector to the smallest resolvable distance in the sample. The value of **d** derived above refers to the object sampled, but for the camera the image source is the focal plane of the microscope optics, which involves the magnification of the optical system used. Thus the smallest resolvable distance for the camera ($\mathbf{d_c}$) is given by the product between **d** and magnification. For a camera attached to a microscope with a Plan-Fluo40 x objective with a NA of 0.8, sampling emission light at 505nm, the object **d** is 385 nm, whereas $\mathbf{d_c}$ value would be 15.4 μm. This would mean that a CCD chip with a pixel size of 15 μm would be enough to detect the smallest resolvable distance. However, most often in practice the interest is to analyse adjacent features. If the pixel size is equal to $\mathbf{d_c}$, then two apposed features in the sample with similar peak light intensities would appear on the camera as continuous. So, what is required then is that the $\mathbf{d_c}$ is sampled by at least two pixels on the detector rather than one (the so-called Nyquist limit). Altogether, a useful equation which comprises all the factors discussed is:

$$(\mathbf{0.61} * \lambda/\mathbf{NA}) * \mathbf{Magnification} = \mathbf{pixelsize} * \mathbf{2} \qquad \text{(v)}$$

which can provide information about either the magnification that will provide maximum optical resolution (for a given pixel size and objective) or about the required properties of a specific objective or CCD chip for a given application.

Comments

1. From this discussion it would appear that a smaller pixel size is preferable, as it gives much better sampling conditions. However, as discussed before, the smaller the pixel size, the smaller the dynamic range of the chip (due to pixel well capacity), so priorities need to be established.
2. When using large pixel-sized cameras and/or objectives of lower magnification and NA, the resolution of the system would be lower than the maximum. One way to increase resolution would be to increase the overall magnification by interposing magnifying tube lenses in the optical system. The danger here is that as more reflection surfaces are added (the magnifying lenses), the light intensity is decreased (flux of photons decreased), affecting the S/N ratio (see above). This could be compensated, if the type of experiments permits, by an increase in the integration time.
3. Also, it must be noted that an increase in magnification reduces the field of view (which can be compensated, at a cost, by an increase in the chip size - from a standard 512 x 512, up to large format 4096 x 4096 pixels).

Temporal resolution (Scan Rate, Readout rate). One of the crucial parameters in determining the temporal resolution is the pixel read rate, which defines the time taken to read the CCD pixel and digitize its output signal. These rates can vary, from camera to camera, between 200 ns (i.e., a readout rate of 5 MHz) and 20 ms (50 Hz). For example, on a CCD chip of 512 x 512 pixels with a readout rate of 1 MHz (1 ms/pixel) a full frame will be read in 262 msec. Thus, the potential frame rate of this chip is 3.8 frames/sec. However, this is the unobtainable ideal, since the acquisition of any frame requires also:

i. exposure;

ii. shutter manipulations (especially for frame transfer cameras (see above);

iii. register shift times (for moving the rows of accumulated charges in the pixels onto the reading registers);

iv. clearing times (required sometimes to clear either the pixel arrays or the serial registers of the accumulated (mainly dark) charges); and

v. serial discard time (when only a subarray of the chip is read and digitized, with discarding the pixels before and after the region of interest).

Of these delays, probably the most important one is the exposure time, and altogether, these delays can add up to 50-100 msec per frame.

The Analog-to Digital converters (ADC) used to digitize the pixel output impose some limitations on the readout rates, especially on chips with a 15- or 16-bit dynamic range (see above) for which the commercially available ADC allow only rates of transfer of up to 500 kHz. Another consideration, apart from the digitization process , is the huge amount of data being generated, especially at the higher bit depth (e.g., a CCD chip like Kodak KAF-6300, with 3072 x 2048 pixels, at 12 bit resolution, produces a frame/image of 9.44 Mbytes!! - see example below), which can impose limits on the ability to move and store such large blocks of data. Finally, more relevant to our discussion on the use of cameras is the fact that an increase in the readout rate is associated, as discussed above, with an increase in the readout noise (an increase in the readout rate from 50 kHz to 1 MHz can be associated with a 2- to 4-fold increase in the noise).

Example The calculation is as follows: 3072 x 2048 (=6.29 Mpixel) x 12 (bit per pixel) = 75.5 Mbits/frame. As 1 byte is defined as 8 bit, then the whole image has 75.5/8 = 9.44 Mbytes

Thus, it is clear that the dynamic range and S/N ratio has to be traded off against temporal resolution. Some modern cameras will allow the user the control of the readout speed ('slow' vs 'fast' scan) and of the 'bit depth' of digitization. For experiments in which temporal resolution is essential, the user can reduce the dynamic range (grey levels) and accept a higher noise level in the images. Another alternative is the pixel binning.

3.2.3 Electronic controls of the camera

Aside from the imaging device, the cameras are provided with a series of essential control circuits designed to enhance the video signal.

Black level defines the lowest level of illumination that is capable of generating a voltage from the camera. For quantitative image analysis it is essential that the cameras provide, instead of an automatic black level circuit (which detects the darkest region of the field and sets it as the black level), a manual control of the black level.

Gain describes the relationship between the input illumination and the output voltage: for the same level of illumination a higher gain will provide a large voltage signal, thus increasing the sensitivity of the camera. The gain is formally reported in terms of electrons per ADU (Analog-to-Digital Unit). A gain of 10 indicates that the system will digitize the signal such that 10 photoelectrons will correspond to 1 ADU (1 count). Usually a gain of 1 x will indicate that the full well capacity of the CCD pixel matches the full range of the digitizer (for a camera with a pixel well capacity of 80,000 electrons at 12-bit digitization, the gain "1x" will mean 20 photoelectrons/ADU). Because gain refers to an amplification process, the reported gain value is actually an inverse of the amplification – thus, is the system gain from 1 to 4, this will mean a 4-fold decrease in the number of photoelectrons which trigger an ADU count, from 20 to 5 electrons/ADU).

Although automatic gain control circuits (AGC) are provided with some of the cameras, allowing the camera to adjust the slope of the relationship by taking into account the darkest and brightest regions of the image, for quantitative image analysis the camera needs to be used at a set gain.

Gamma correction. Gamma (g) is a number that represents the relationship between the input intensity and the output voltage, where the signal output varies with the gamma power of the input intensity. When g is less than 1 the sensitivity of the imaging device decreases as the light intensity increases, in a non-linear fashion (with a higher gain in the dark regions of the image than in the light areas). For image analysis purposes, the use of a gamma value of 1 (ie, linear relationship between input and output) is essential.

3.2.4. Detection Modes

In the following paragraphs we will try to integrate the theoretical concepts discussed, and provide a practical guide for using cameras for different types of applications.

High Sensitivity Mode. This setting aims at capturing the smallest increases in light output from the sample, in conditions of initial low-light intensity (e.g., Ca^{2+} puffs in resting conditions, with moderate dye loading). To achieve this mode, the gain of the system (ie, the number of photons triggering an Analog-to-Digital Unit (ADU)) must be set in such a way as that the entire bit depth (e.g., 4096 levels for a 12-bit system) covers only 1/3-1/4 of the full well capacity of a pixel. The major drawbacks of this setting are i) the rapid saturation (any significant event will saturate the sensor) and ii) the fact that it requires a very 'quiet' background.

High Dynamic Range Mode. This setting is required when comparisons between signals within regions with very different intensity levels are assessed (e.g., a Ca^{2+} response to stimulation in a neuron, comparing response in the soma, which will have, due to its larger volume, a much larger resting signal than the neurite areas). In this mode the gain of the system will be set in such a way that the full well capacity will span the entire bit depth of the Analog-to-Digital (ADC) converter (i.e., for a 12-bit system, the 4095 value will be reached at single pixel full well capacity). Clearly, the higher the bit depth of the camera's ADC, the higher the potential dynamic range of the image recording, but at the expense of missing subtler, smaller events. This mode, as well as the previous one, takes full advantage of the spatial resolution of the camera.

Pixel Binning Mode. This mode of capturing is frequently used and aims at improving either the S/N ratio of the sensor or the temporal resolution (or both), but at the expense of sacrificing spatial resolution. An increase in the S/N ratio is beneficial, for example, when high precision measurements are required for small intensity changes against a bright background (e.g., use of bright dyes, changes in $[Ca^{2+}]_i$ in conditions when the resting $[Ca^{2+}]_i$ is already high).

Binning is the process of combining electronic charges from adjacent pixels in a CCD chip during readout into a single "superpixel", a process which takes place before digitization. Almost all modern imaging cameras will have on-chip circuitry for control of the parallel (transfer) registers and the horizontal (serial, readout) register to allow binning. Let us take the example of a 2 x 2 binning. In this case, after the end of the exposure time, the charges from two, instead of one, parallel registers are transferred to the serial, readout register. Further, two, rather than one, pixels of this register are transferred towards the output node (digitizer). Thus the output node receives the charge equivalent to a 2x2 (4) pixels on the sensor are digitized as one reading ("superpixel"). Software control permits various binning protocols (from 2x2 to 8x8 or more).

While the increase in temporal resolution through binning is evident (less pixels to be read/digitized), the effect on the S/N requires some clarification. The main reason is a significant reduction (the larger the binning, the larger the reduction) of the readout noise. This type of noise (see above) is added during each readout event, and added to each pixel read. However, in binning mode, there is only one "super" pixel read, while the signal is 4 times larger, thus the S/N is increased, potentially, 4 times. (for a 3 x 3 binning the increase is 9 times, and so forth). However, the increase is not exactly 4 times, because of the other sources of noise, which are pixel-based. For example, the dark current noise is not reduced by binning, since each pixel in the "superpixel" will carry this statistical noise. A proper cooling of the CCD will reduce, as discussed above, this source of noise.

Binning will also have a positive impact on the dynamic range. As presented above, the dynamic range is effectively described by the ratio between the full well capacity of the sensor and the readout noise. The important point which needs to be mentioned here is that, in modern cameras, the well capacities are increased along the pathway of charge transfer: the serial register has typically a well capacity double that of the parallel registers, and the output node double the capacity of the serial register. At the same time, the well capacity (electronic charge) of the output node, which collects the electronic charges at the end of the horizontal register from each pixel presented to it and making it ready for the ADC, also sets the limits of the efficiency of the binning procedure, when the lighting conditions have been taken into account. In low-light level, binning increases the S/N ratio of the image and its dynamic range and, since there are only relatively few photoelectrons produced by each pixel, there is no danger of saturating the output node, even at a high binning level. Thus, binning neighbouring pixels on the CCD array may allow one to reach a photon-limited signal (ie, an image with a good S/N ratio) with a shorter integration time, at the expense of spatial resolution. In contrast, at much higher levels of illumination, one might be tempted to use binning for an increase in the temporal resolution, but an excessive binning might induce saturation at the output node, since each pixel on the sensor will contribute a significant amount of photoelectrons to the "superpixel" read.

3.2.5. Increase of camera sensitivity – Image intensifiers

Measuring intracellular Ca^{2+} with camera technology means , in most cases, working with fluorescence, in low light levels, demanding very sensitive systems. One way to increase sensitivity is by increasing the gain of the camera. Another one is to increase the integration time on the chip, but this method is severely limited by the dynamic nature of the process (Ca^{2+} signalling) studied. Finally, one of the most effective means of increasing the sensitivity of a camera (up to 10,000 times or more) is the use of an image intensifier.

The modern image intensifier consists, generally, of three units: a photocathode, a microchannel plate (MCP) and a phosphor screen enclosed in a high-vacuum environment (to prevent interaction between electrons and gas atoms, which will generate positive charges, rather than negative charges, which will be accelerated towards the photocathode and potentially distort the electron emission at that site). The consists of a thin layer of material onto which the image is focused and which generates electrons following the photon impact. By choosing the appropriate material, the spectral sensitivity can vary from ultra violet to infrared. For many materials, the QE exceeds 10%. The electrons generated by the photocathode are accelerated by a strong electric field towards the MCP. Depending on the chosen voltage at the photo cathode, the electrons are either accelerated towards the MCP (negative voltage) or remain in the photo cathode (positive voltage). Consequently, the image intensifier can be used as an extremely fast (voltage changes can be implemented within 5 nsec) and efficient electronic shutter which can be used for picking up faint pulsed signals against a bright continuous signal or image rapid (nanosecond time scale) changes of transient events.

The **microplate channel** is made out of leaded glass and includes 1-10 million little channels with a characteristic diameter of 10 μm and a length of approx. 0.5 mm. Each channel represents a photo multiplier, and each time an electron hits the wall of the channel it will generate further electrons by secondary emission. The degree of intensification depends mainly on two parameters: a) the length and diameter of the channels (increasing the number of hits) and b) the applied voltage. Therefore, the degree of intensification can be influenced by setting the acceleration voltage (gain). Finally, the electrons leaving the MCP are accelerated, at a voltage of several kV, towards the anode, which is a **phosphor screen** on which they generate the photon emission. The main factors in choosing a proper phosphor screen are the energy efficiency and the luminescence. Depending on the phosphor used, the luminescence can vary from some nanoseconds to several seconds and must be carefully chosen for each application.

The final significant step is to link the image produced on the phosphorus screen onto the CCD chip. This is usually implemented either through lens coupling or by tapered fiber optic coupling. The choice has important implications and the potential users should be aware of these.

a. By choosing lenses as the coupling system, one obtains a picture that is nearly free of distortion and, also, one can freely choose the scale, therefore adjusting the resolution of the image intensifier and the CCD sensor to each other. Especially for smaller CCD chips, the picture on the phosphor screen, which has only limited resolution, can be scaled down and projected onto the CCD sensor. Consequently, the picture obtains a spatial resolution that is adjusted to the size of the pixels of the CCD sensor. Another advantage is that the intensifier can be removed from the system, and, when the light conditions of the samples allow, the CCD camera can be used on its own. However, these advantages are severely counterbalanced by the comparatively very low efficiency, which for lenses with a characteristic aperture of 2.0 has a value of only 1.5 percent approximately.

b. Coupling with a fiber optic taper is essential when a maximum efficiency of the light transmission is required. By choosing the proper materials, one can reach a numerical aperture of 1, thus allowing the transmission of all the light intensity from the phosphor screen onto the CCD sensor. The fiber optics can be calibrated more easily than the lens optics, since distortions relative to the optical axis which might occur hardly influence the quality of the coupling. A possible disadvantage is the presence of some optical irregularities resulting from the production, but these can be compensated for.

3.3. Real time confocal microscopy

We are discussing confocal microscopes in the section dedicated to detection systems, mainly because these instruments are normally available as a complete, turn-on package. As it will become immediately apparent, in reality 'confocal' refers in principal to a particular method of illumination of the biological specimen. In the last decade real-time confocal microscopy has been established as a widespread tool for measuring intracellular calcium in living cells. From the practical point of view, the main advantage of the confocal microscope is its ability to collect signals from restricted volumes, thus giving a possibility of "optical sectioning" of the biological object. 'Confocal' just means 'same focus, and, by definition an object and its image are always confocal. However, when an object is seen through a conventional fluorescence objective, the signal produced comes not only from the focal plane of the image but from the full thickness of the specimen. It is the capacity of the 'confocal' microscope arrangement which allows the collection of an image in a very narrow focal plane, with the exclusion of the light originating in other planes outside the focal plane, which in effect provides the 'optical sectioning. This could be achieved by employing two basic principles (Fig. 6), namely (i) restricting the signal collection (conventional confocal microscopes) or (ii) restricting the excitation of the fluorophore (multiple-photon confocal microscopes). For an in-depth treatment of confocal microscopy principles and uses, the reader can refer to comprehensive volume, published recently (Paddock, 1998).

In conventional confocal microscopy, a single point in the specimen is illuminated either through a spinning Nipkow disk situated in a conjugate focal plane to the specimen or, much more commonly, through focussing a parallel laser beam. The reduction in the 'out-of-focus' light contamination comes from two sources - first, with a single point illuminated, the illumination intensity rapidly falls off above and below the focus plane, as the beam is either converging (before the plane) or diverging (beyond the plane) and, secondly, from the insertion of a pinhole in front of the detection system, which allows only the light coming from the focal plane to the detector. The multiple-photon systems, instead, use an energy summation of two or more photons, which emerges when the latter meet, which is

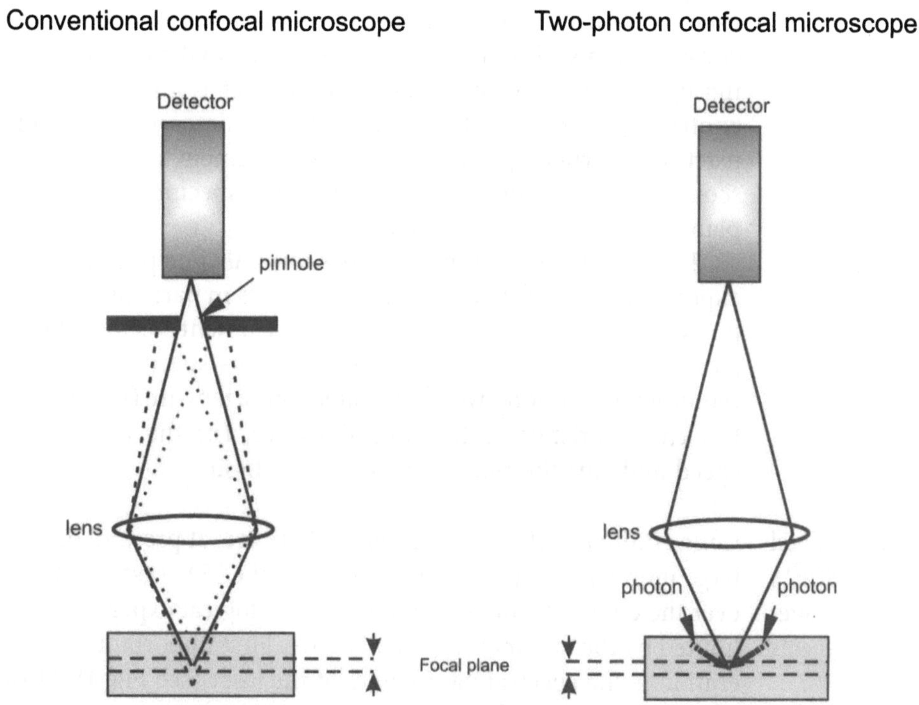

Fig. 6. Principles of confocal microscopy. See text for details.

most probably only at the focal plane. As the summation energy depends on the square of the incident light intensity, the excitation of fluorophore is limited to a very narrow point.

These descriptions of principles apply for single point. To record an image, the laser beam has to scan the field of view, which is achieved by a system of motorized rotating mirror, which sequentially scans a horizontal beam across the specimen. This arrangement gives the instruments their general name of confocal laser scanning microscopes (CLSM). The large-scale introduction of confocal microscopes in biological research relied on a number of elements. Compared with a normal fluorescence microscope, the amount of light that is seen in the final image is drastically reduced by the pinhole, up to 90-95% (depending on the size of the slit opening). To overcome this, two components had to be incorporated: a very strong illumination source, ie, lasers (see previous sections) and highly sensitive detectors (PMTs). In addition, essential for the development of modern confocal microscopes was the increase in computational power of desktop computers. The computer is used to control the system of scanning mirrors as well as synchronizing this with the collection of data from the PMTs, building up the final image. The same computer is also capable of controlling, for some types of confocal microscopes, mainly used for morphological studies, the movement of the microscope stage, resulting in sectioning of the specimen, one focal plane after another, in the z plane.

The use of confocal microscopes for real-time physiological experiments, and for $[Ca^{2+}]_i$ measurements in particular, as opposed to the other major use of this instrument for morphological and immunocytochemical studies, sets specific hardware requirements, among which the most important are (i) the spectral characteristics of the optical source, (ii) the acquisition speed and (iii) the photodamage of the tissue.

3.3.1. Confocal $[Ca^{2+}]_i$ measurements

Conventional confocal microscopy. There are, at present, quite a large number of suppliers of conventional CLSM systems, however the choice of one suitable for physiological experiments is limited by the factors mentioned above. First important consideration is the spectral parameters of the laser (see Fig. 4). Most of CLSM use Argon lasers with a main excitation wavelength of 488 nm, restricting thus the choice of Ca^{2+} indicators to non-ratiometric dyes only. This in essence limits the experiment to qualitative recordings of intracellular Ca^{2+} dynamics, rather than measurements of actual $[Ca^{2+}]_i$ values. Alternatively, certain CLSM systems can be equipped (at a substantial additional price) with UV lasers, which allows the use of double emission ratiometric dye indo-1. Unfortunately, illumination with UV lasers delivers very high levels of energy into the object (and this energy is dispersed through the whole preparation), which leads to a rapid bleaching of the dye and causing a substantial photodamage. The second important parameter is the acquisition speed of the system. Most of commercially available CLSM are relatively slow, acquiring a single image within 1 - 3 seconds, which very much limits their use for real-time $[Ca^{2+}]_I$ measurements. Certain systems, though have a reasonable acquisition speed, allowing the capturing of 15 - 30 images per second (e.g. Noran Odyssey XL). The problem of low acquisition speed can be circumvented by using a particular feature of CLSM limiting the object scanning to a single line. This particular line scanning could be done at a great speed (50 - 100Hz) at almost each model available, and may deliver relevant information about Ca^{2+} dynamics in certain cells/cellular compartments.

Multiple-photon confocal microscopy. The multiple-photon confocal microscopy (MPCLSM) overcomes most of the limitations of conventional CLSMs. According to its principle, the MPCLSM uses the energy of pulsed infrared laser to excite either UV or visible-light fluorophores by simultaneous absorption of 2 or more infrared photons. In this case, each infra-red photon carries only a fraction (1/2 or 1/3) of energy required to excite the fluorescent probe. When the infrared photons collide within the focal plane they provide the fluorophor with the appropriate excitation. This excitation, though, occurs strictly within the focal volume when several excitation photons can be absorbed simultaneously. This has the following advantages: (i) any fluorescent probe (either UV or non UV) can be used; (ii) the photobleaching and photodamage are limited to the focal plane only, as the rest of the preparation does not absorb energy necessary for dye excitation; (iii) the MPCLSM has a deeper penetration power: the infrared photons can travel through the biological objects far longer. This offers a particular advantage while working with in situ preparations as the cells lying within the tissue can be studied. In fact, the use of multiple-photon technique can be extended for a whole tissue in a living animal, such as studying the calcium dynamics in nerve cells in a whole brain (Svoboda et al., 1997). All these advantages though, come at the price of approximately two-fold decreases in a lateral resolution of MPCLSM.

Further Information

At the present time, the Internet, through the WorldWideWeb, is becoming an established repository of information. All the companies mentioned in this article have a Web presence, in which details of the range of products available and, sometimes, technical details can be found. Since sometimes the name of the company differs from the domain name of the Web site, the best strategy is the use of one of the numerous search engines. Amongst these we can mention: AltaVista (http://www.altavista.com/); Inforseek (http://infoseek.go.com/) or Google http://www.google.com/). The search can be done either for the company name, or for one of the specific products of interest. The use of search engines also has the advantage that it will provide several other 'hits' (hundreds to thousands), of which some could be of interest (usually the 'hits' are displayed in the order of their relevance to the search string requested). Amongst these secondary hits, one should note the existence of several personal pages (pages produced by individual researchers or by their labs), dedicated to one or another aspect of fluorescence microscopy, which contain both technical information and several very useful links to other sites of interest, allowing one to start 'surfing' the Net, on the crest of a wave of interest. Just to give one example, Lance Ladic at Univ. British Columbia, Canada http://www.cs.ubc.ca/spider/ladic/confocal.html) maintains a very useful page for confocal microscopy.

References

Garaschuk O, Yaari Y, Konnerth A (1997) Release and sequestration of calcium by ryanodine-sensitive stores in rat hippocampal neurones. J Physiol (Lond) 502: 13-30

Grynkiewicz G, Poenie M, Tsien RY (1985) A new generation of Ca^{2+} indicators with greatly improved fluorescence properties. Journal of Biological Chemistry 260: 3440-3450

Hofer, AM Schulz I (1996) Quantification of intraluminal free [Ca] in the agonist-sensitive internal calcium store using compartmentalized fluorescent indicators: some considerations. Cell Calcium 20: 235-42

Kirischuk S, Verkhratsky A (1996) $[Ca^{2+}]_i$ recordings from neural cells in acutely isolated cerebellar slices employing differential loading of the membrane-permeant form of the calcium indicator fura-2. Pfluegers Arch 431: 977-83

Neher E (1995) The use of fura-2 for estimating Ca buffers and Ca fluxes. Neuropharmacology 34(11): 1423-42.

Paddock SW (Ed) (1998) Confocal Microscopy Methods and Protocols (Methods in Molecular Biology Vol 122) Humana Press 464 pp.

Svoboda K, Denk W, Kleinfeld D, Tank DW (1997) In vivo dendritic calcium dynamics in neocortical pyramidal neurons. Nature 385(6612): 161-5.

Simple Principles of Imaging – Image Files, Image Processing and Image Analysis

EMIL C. TOESCU

Introduction

More often than not, the capturing of an image, or of a series of images, is only the beginning of a long, hard journey of extracting the information and, on some occasions, of preparing the image(s) for publication. The array of image analysis and manipulation programs and suites is huge and increases continuously, as the computer power and memory increases all the time allowing the transformation of the computer from a text-based machine (in good old "DOS times") to an image-based system.

1. Basics of an Image

First, one needs to distinguish between the two fundamental types of digital images: a) the vector-based images (such as a Corel drawing), in which the information is stored as vectors (starting point, direction, angle, length, fill, etc) and b) the **bit-mapped** images, in which the image is composed of rows after rows of pixels (**pic**ture **el**ement). For imaging, the bitmapped image is the final product, and the pixel on the screen is generally equivalent to the pixel on the CCD chip or to the individual point scanned by the confocal instrument. The standard, medium-format imaging sensors of today have sizes of 512 x 512 or 640 x 480 pixels, although larger, mega-pixels chips are becoming available (around 1k x 1k).

Emil C. Toescu, Birmingham University Medical School, Dept. Physiology, Division of Medical Sciences, Birmingham, B15 2TT, UK (*phone* +44-121-414-6927; *fax* +44-121-414-6924; *e-mail* e.c.toescu@bham.ac.uk)

Another essential parameter which describes the richness of information contained in a digital image is the **pixel depth**. At the most simple, a pixel can be 1-bit deep – that is the pixel is either white ('on') or black ('off'), resulting in a binary image, very useful for performing some types of morphological analysis of images. However, this kind of image is too poor in information and is almost never produced by the image detector. The minimal standard of image output has been, for quite some time, the 8-bit image (8 bit = 1 Byte). In this image, each individual pixel has a value between 0 and 255 (2^8=256). The size of such a 512 x 512 image would be: 512 x 512 x 8=2.1 million bits or 262 kBytes. Such an image is also known as a monochrome image, with all the 256 levels representing in fact grey levels, between black (0) and white (255). Through subsequent image segmentation and application of look-up tables (LUT) (see below), this image can be then pseudo-coloured, to better convey an impression of the grey level distribution.

However, when using imaging arrangements with multiband filters (see previous chapter), which provide information from the various regions of the visible (coloured) spectrum, one will need to use coloured CCD chips. In these cameras and in the resulting images, each digital colour image pixel carries separate 8-bit information about each of the three main colours: red, green and blue (RGB image), resulting in a final 24-bit image. A pixel encoding (255,255,0) will be represented on a screen as bright yellow (red and green fully saturated, no blue), while a (255, 255, 255) is white and (0,0,0) is black. Overall, each pixel in this 24-bit image can display one of 16,7 million colours (256^3),and the size of a 512 x 512 image will be 786 kBytes.

Returning to the monochrome images, although 8-bits images may be sufficient for displaying or printing, the limited digital information (only 256 levels) may be insufficient, especially when high measurement accuracy is required or when the images contain structures with very different levels of fluorescence intensity (such as a whole neuron, in which soma and dendrites have very different intensity levels). For these applications the use of more modern, and expensive, digital cameras, with pixel depth of 12-16 bits/pixel are recommended. A 12-bit depth will provide 4096 grey levels (2^{12}), and a 16-bit, 65,536 grey levels.

2. Image File Formats

An image file contains, basically, a specified number for each pixel/location in the image and the data is organised in rows and columns. There are many image file formats, and one important difference between them is the structure of the file header, where details about the organisation of the file (image size, number of rows/columns, time stamps, compression algorithms, etc) is stored. For display, a program accessing this file needs to be able to read this information, and its (in)ability to read file headers creates much anguish and many lost hours.

There are many types of image files, and the first distinction must be made between the proprietary image formats used by the various image capture/analysis suites available and the more general image formats used for transferring image files between programs/computers/web locations.

The first category of file formats includes the *.AIF (Axon Image Format), BioRad's *.PIC files, Universal Imaging Corporation *.IMG files, Imporvision's *.LIFF (on Mac platforms), etc. The reason for their existence is simply that each of these image suites are capable and required to perform complicated manipulations of image files, in real-time, and/or building image stacks which all require specific software code writing and each house developed ways of maximizing the efficiency of files manipulations. To a great extent all these specific file formats are 'black boxes' for the users, who just have to concentrate on the actual experimental protocol rather than on the details of file manipulations. However, a discussion of the general image file formats is important, because this is how the files are generally transferred between programs, when the image analysis facilities in the capturing program are insufficient. An important characteristic of most of these file formats is that they are generally platform-independent (i.e., - they can be used on PCs, Macs, Unix, Linux, etc). A very detailed discussion of these file formats (updated to 1995) can be found elsewhere (Brown & Shepherd, 1995).

2.1. TIFF files

Probably the standard general image file format is the **TIFF** file (Tagged-Image File Format, extension: *.tif or *.tiff) and almost all current image analysis programs and image capture suites will support (import/export functions) this format. Standard TIFF files supports 1,4,8 and 32 bits greyscale and 24 bits RGB images and the way the format was designed allows a very large number of possibilities for saving an image. Details of how the image is actually organised is kept in the image header, which can contain any of up to 70 different tags, each tag containing information about the image width, height, colour table, etc. The file format is flexible and extensible, and additional features can be added by designing additional tags (for the latest TIFF 6.0 specification check at ftp://ftp.adobe.-com/pub/adobe/devrelations/devtechnotes/pdffiles/tiff6.pdf). However, this flexibility comes at a price, as no image analysis application can claim to support all TIFF variations (tags), and sometimes obscure TIFF tags cannot be read. Another important feature of the TIFF files, and probably the most important source of complications when reading TIFF files, is that the TIFF files can also be compressed (a lossless compression, see below) by a variety of algorithms, and not all are supported by all imaging programs. The compression methods include the LZW (Lempel-Ziv-Welch) format (supporting 4,8,24 bits/pixel), or the Huffman and PackBits methods (supporting only 1 bit/pixel, meaning that they are effective only on binary images, not on continuous tone images). As a practical tip - if an application is having a problem reading compressed TIFF files, try re-saving the file without compression.

Understanding image file compression: imagine a pixel string as follows: 25, 25, 26, 29,29, 29, 25, 25. A lossless compression algorithm will show as 25(2), 26, 29(3), 25(2), saving some space (the more uniform the image, the higher the compression), whereas other lossy compression methods will result in 25(3), 29(3) 25 (2), or, even more compressed, as 27(8). Lossless compression results in smaller file sizes without loss of image quality. GIFs, PNGs, and TIFs with LZW compression are all examples of lossless compression. Lossy compression generally results in greater reduction of file sizes, but also in increased loss of image quality. JPGs are a type of lossy formats.

Despite its complexities, the TIFF file format is one of the best for transferring bitmaps across platforms, as it is flexible enough to allow virtually any image to be encoded in binary form without losing any of its attributes, including mask information (alpha channel), layer(s) information or attached text information. In principle, the files should be saved as uncompressed TIFs, and, if required, the recommended compression routine used should be the LZW.

2.2. BMP files

Another common file format is the **BMP** format (Windows **B**it-**M**a**P**, extension *.bmp), which is native to the Windows environment and can thus be displayed on any type of display device (monitor), without any need for 'translation'. This file format supports 1,4,8 and 24-bit colours, and thus is adequate for storing imaging files (either greyscale 8-bit or 24-bit RGB). However, for compression, the BMP files are restricted to 4 or 8 bits/pixel image formats using the RLE (run-length encoding) algorithm (becoming and *.rle file).

2.3. Other common file formats

Other image files formats, relatively common for image manipulation packages, but less used for actual handling of imaging files produced by the major imaging suites, are **EPS** (Encapsulated PostScript) files (extension *.eps or *.epi), which is widely used and accepted by electronic publishing systems; **TGA** (Truevision Targa format) files (common extension *.tga), mainly used for images grabbed from Truevision video boards; **PCD** (PhotoCD; extension *.pcd) produced by Kodak; **PSD** (PhotoShop Adobe; extension *.psd) as the proprietary file format (native) for Adobe PhotoShop program.; **RAW** format containing the raw pixel data, un-coded.

With the expansion of the Internet, it is becoming more and more common to want to have images displayed on various sites. All the file formats discussed up till now result in relatively large files and the browsers do not always support the respective formats for decompression. Two other image file formats are now established for use on the Internet, and an important principle about them is that both save directly in a compressed format.

2.4. GIF files

GIF (Graphics Interchange Format; extension *.gif) is a common file format, but it is not recommended for use with primary digital microscopy images because it only supports image files with less than 8-bit depth. The idea behind designing the GIF format was to create the smallest possible image file for uploading and downloading from electronic Bulletin Board Systems (BBS), thus producing a highly compressed format that minimizes file transfer time over phone lines.

The GIF format uses an encoding method (based on the LZW method) referred to as interlacing, i.e., the image is saved by using four passes instead of just one. On each pass, certain lines of the image are saved to the file. If the program decoding a GIF file displays the image as it is decoded, the user will be able to see the four passes of the decoding cycle. This will allow the user to get a good idea of what the image will look like before the full image is decoded. Most communication programs for Internet allow the user to download GIF files and view them as they are downloaded. If the image is interlaced, the user will be able to decide if the image is one they like/need before half of the download is complete, and abort, if necessary, the full download.

GIF files, like the TIFF files, may contain multiple images, together with other information concerning overlays and text to be displayed with the image. Another attractive property of the GIF files is the capacity for supporting animation – multiple images are saved in the same single file and can be displayed frame-by-frame in a mini-film (various image manipulation packages (e.g., PaintShop Pro v5.0) support the buildup of animation sequences.

2.5. JPEG files

The JPEG format (pronounced 'jay-peg', Joint Photographic Experts Group, extensions *.jpg, *jif) was developed by C-Cube Microsystems to provide an efficient method of storing images of higher pixel depth (normally, 24-bit) which are characterized by numerous subtle variations in colour (16 million colours; see above). The crucial feature of this format is that the storing/compression is a lossy procedure (see above), which sacrifices some image data in order to achieve significant compression ratios. A decompressed JPEG image rarely matches the original exactly, but very often the differences are so minor that they are barely detectable, if at all. The JPEG image compression is a complex process that frequently requires a hardware assist to achieve acceptable performance. First, the image is tiled into blocks that measure 8 pixels to a side. Each block is then compressed separately, in three stages. At the first stage, the image is divided into blocks of 8-by-8 pixels, and, using a mathematical conversion (a discrete cosine transform (DCT)), the blocks are converted into an 8-by-8 matrix of amplitude values, representing different frequencies (or rates of colour change) in the block. At stage two, the values in each amplitude matrix are divided by the values in a quantization matrix, which stores all the amplitude values in the image. This quantization matrix can be biased by the operator to filter out amplitudes that are less important to the overall appearance of the image. At the third and final stage, the quantized amplitude matrix is compressed using a lossless compression algorithm.

The lossy part of the JPEG compression is the second stage. The higher the values used in the quantization matrix for filtering out amplitudes, the greater the amount of information discarded from the image, and the more tightly the image is compressed. The trade-off is that higher quantization values result in poorer image quality. The optimal quality factor – the one that exhibits the best balance between compression ratio and image quality - is different for every image and is usually found only through trial and error. The compression factor in JPEG varies from 2 to 100, and usually, even for digital microscopy files, compressions of up to 10-15 do not introduce visible artifacts, while TIFF files can usually only achieve a 2-3 fold compression. However, the crucial point is that JPEG saving always loses data, so it can't be used for storage of raw data, but only for display purposes. Another very important practical point about handling the JPEGs is that, as saving a JPEG file means compression, and loss of information, any saving of the image file, as an intermediate step, will reduce its information.

2.6. Newer file formats

Finally, the discussion about image file formats must mention the new emerging formats, which very probably, will soon replace today's standards. **DICOM** (**Di**gital Imaging and **Com**munications in Medicine); extension *.dcm) is an image format standard which has been developed for formatting and exchanging medical images and associated information (e.g., X-rays or CT-scans). As DICOM supports up to 16 bits/pixel for monochrome images and 24 bits/pixel for RGB pictures it is well endowed to support future applications in digital microscopy. DICOM files allow multiple images and/or cine sequences and the images can be compression with either lossy and lossless algorithms, providing further flexibility. Another format which is developing very rapidly as a standard is the **PNG** (**P**ortable **N**etwork **G**raphics, extension *.png) format, which supports 16-bit monochrome and 48-bit colour images and was designed to replace the GIF format using also a lossless compression (further details can be found at http://www.cdrom.com/pub/png).

3. Basic Image Processing

For all those involved in digital imaging, it is a fact of life that the raw images captured during the experiment are just that – raw images, and require a significant amount of work for the extraction of information. This effort could be divided, for the sake of organising this presentation, into several stages. The **image capture** stage (Fig. 1) has been discussed in some detail in the previous chapter, whereas the file formats adequate for **image display** (Fig. 4) have been presented above. During the rest of this chapter, some of the general principles involved in performing the other two image manipulation stages: **image processing** (Fig. 2) and **image analysis** (Fig. 3) will be presented. For an

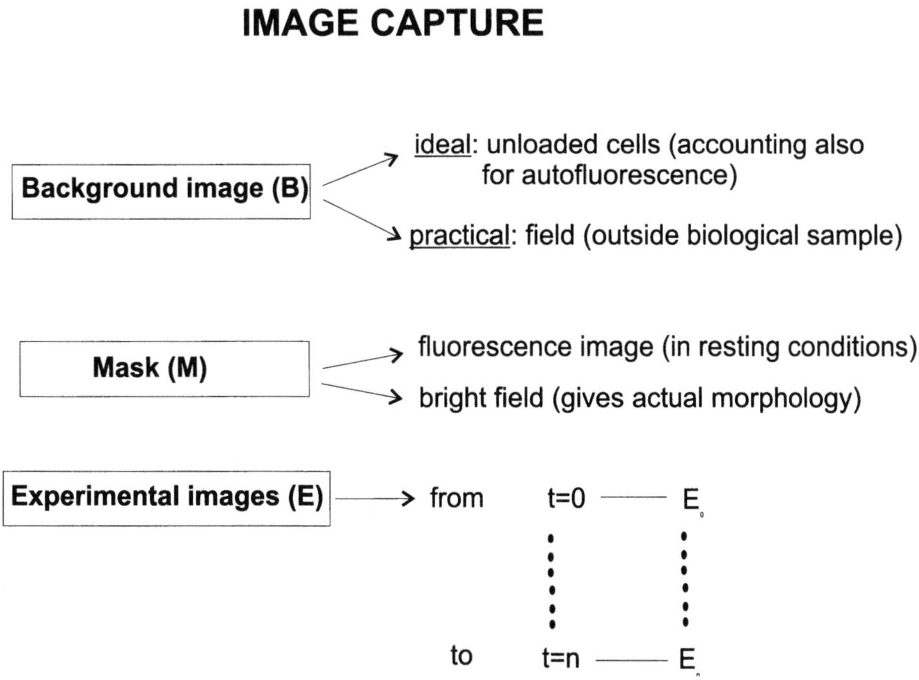

Fig. 1. Steps required for capturing a stack of images during a Ca^{2+} measurement experiment. The capturing of a background image is particularly important for the subsequent processing of the image stack. The number of images in the stack (E_0 to E_n) can vary from ten to hundreds, depending mainly on the memory capacity of the computer and its ability to stream this information to the hard disk.

in-depth discussion, the reader is referred to one of the many books available on image processing and analysis (e.g., Glasbey & Horgan, 1995; Russ, 1995).

3.1 Image Processing

3.1.1. Contrast enhancement

In order to obtain the greatest amount of precision from an image analysis system, it is important to obtain the maximum grey level resolution possible – in other words, that the image will use as many of the grey levels available as possible – in effect, a maximal intensity resolution. The grey level histogram, available in any image processing packages, provide a good representation of the contrast in the image. The histogram consists of a plot of

IMAGE PROCESSING

1. **Shading and Background correction** (Batch operation)

$$(E_0 \ldots E_n) - (B) = (E'_0 \ldots E'_n)$$

2. **Masking** (Batch operation)

$$(E'_0 \ldots E'_n) - (M) = (E''_0 \ldots E''_n)$$

3. **Ratioing** (Batch operation)

   a. <u>single wavelength dyes</u>

   $E'_0 \ldots E'_i$ (resting conditions) = E'_0 (average); then
   $(E'_{i+1} \ldots E'_n) / E'_0$ - to obtain ratio images

   b. <u>ratiometric dyes</u>

   - requires previous calibration of system
   - uses LUT - **Image segmentation**

4. **Further processing**

   Spatial filters - Sharpening filters
   - Smoothing filters

Fig. 2. Steps involved in the initial processing of the experimental images. The first step, and the second one if required, are performed in batch mode (i.e., the subtraction operation is done on all the files in the image stack). The third step, which is also a batch operation, differs slightly from the two conditions described, a) and b). For single wavelength dyes (a), it is usually the same averaged image, obtained in resting conditions, which is the denominator for the whole of the stack. For ratiometric dyes (b), the ratioing takes place at each individual time point (e.g., for a fura-2 experiment, at time t=t, the 340 nm image is divided by the 380 nm image) and the resulting image is then compared with the calibrated LUT tables.

the number of pixels in the image that are at each grey level (or range of grey levels). Maximal intensity resolution (maximal contrast) is achieved when the pixels are distributed across the whole grey level range. Consider an overexposed image, in which the grey level histogram shows that the grey levels in the image stretch only between 100 and 255. To correct this, and reveal the details of the image it is recommended

IMAGE ANALYSIS

1. Measuring $[Ca^{2+}]_i$ values (Area analysis)

i) Define ROI on the ratio images

ii) Measure **mean** pixel intensity (equivalent to $[Ca^{2+}]_i$ values) in each ROI against time

2. Measuring $[Ca^{2+}]_i$ values (Line scan)

i) Define line of interest in one of the initial images (in confocal microscopy this is done before image capturing)

ii) Plot pixel intensity (equivalent to $[Ca^{2+}]_i$ values) across the line and repeat along the time line.

Fig. 3. When only $[Ca^{2+}]_i$ measurements are performed on the stack of images, and no morphological information is required, then the operation is restricted to defining the regions of interest (ROIs), usually cell bodies or some subcellular compartments, and asking the computer to perform either a density measurement over the area of interest (1) or a plot of the individual grey level values of pixels situated across a line of interest.

to perform a histogram stretch. This operation samples the image to obtain the lowest and the highest grey levels and then builds up a look-up table with the following formula to each pixel:
NewGrayLevel = 255 x (OldGreyLevel - MinimumGreyLevel)/ (MaximumGreyLevel - MinimumGreyLevel)

3.1.2. Shading correction

Shading is a distortion in the image, due either to small misalignments in the optical path (especially of the condenser) or to a lack of uniform sensitivity of the imaging detector to light. The extent of the problem can be usually assessed by examining a blank, uniformly illuminated field. The image captured on this occasion would provide also the means of correcting for the shading, by applying a procedure called background subtrac-

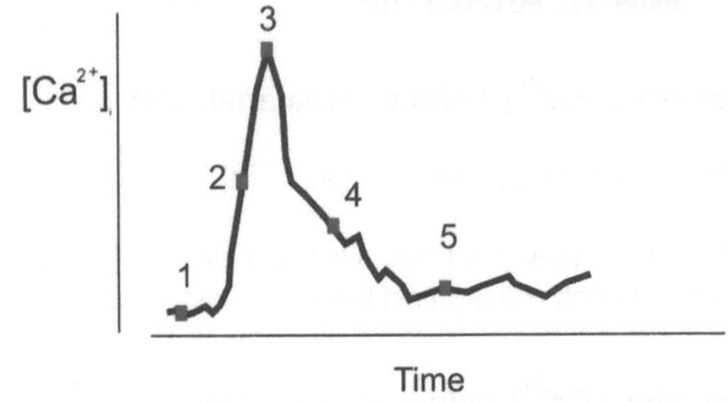

Fig. 4. For display purposes (printed publications or slide presentations or display on web sites), the images presenting the features of interest are first selected, and then cosmetic changes are performed, preferably using a dedicated image processing package, and which include the final montage of the image. The type of file format used for saving the images depends on the subsequent use of the image.

tion, in which the pixel value in the background image (collected on the blank field) is subtracted from the pixel value of the experimental image. This background subtraction should be a routine operation for each individual imaging experiment.

3.1.3. Spatial filters

Filtering in the spatial domain, as opposed to applying filters for the colour information, is an extensive set of mathematical operations (image convolutions) that serve to alter the grey levels of a particular pixel based on the grey level values of the neighbouring pixels. These neighbouring pixel values are given various numerical weights depending on the operation to be performed. The range of effects and applications of these type of filters is wide, including the reduction of noise (smoothing), enhancement of detail (sharpening), detection of edges of various objects present in the image. For performing an image convolution a kernel is created, consisting of a grid of pixel values, centered by the pixel of current interest. The kernel is usually an odd number value, so as to give mirror symmetry around the central pixel: e.g., 3 x 3, 5 x 5, etc. Associated with the convolution kernel is a matching array called the convolution mask, or filter, which contains the weighting factors to the pixels in the kernel.

Application of the spatial convolution proceeds by first multiplying the value of each pixel element in the kernel by its corresponding weighted value, then adding all the resulting values, and assigning this value as the new value of the central pixel – in the following example, the initial value of 48 will be replaced by 79:

28	34	35
30	(48)	52
38	69	67

Kernel

X

-1	-1	-1
-1	9	-1
-1	-1	-1

Mask

=

-28	-34	-35
-30	432	-52
-38	-69	-67

Result

→ **79**

New value

The kernel is then moved to the next pixel (in our example, 52) and the mask applied again, the process continuing for all the pixels in the image.

When an image contains a considerable amount of fine detail, resulting in many sudden transitions of intensity over short spatial distances, the image is said to have a large amount of high-frequency information. When most of the grey scale gradations in the image are small or occur over long distances, the image has low-frequency components. The filters may selectively enhance or diminish the high-frequency components. High-pass filters (such as a Sharpen filter (see below) will enhance the detail by selectively increasing the grayscale value differences. A low-pass filter (such as Smoothing filter) will decrease these high frequency components by reducing the intensity value transitions.

Sharpening Filter. This filter's function is to increase the high-frequency component, while leaving the low-frequency relatively unaffected. As a result the fine details in the image will be enhanced, and also it can, to some extent, reduce the effects of poor focus or haziness in the image. It is important to understand that this operation cannot 'bring in' detail which was not in the original raw image, it can only 'bring it out'. The operation is achieved by using a mask that weights the central pixel heavily, while decreasing the contribution from neighbouring pixels. An illustration of such a 3x3 mask is illustrated in the example used above – the higher the central weight the higher the amplification of the high frequency elements. Larger filters (5x5 or 7x7) can be applied and the weight applied in the surroundings can be set in such a way as to highlight the details even further by decreasing the contribution of neighbouring pixels. Image manipulation programs, such as PainstShop Pro v5.0 (shareware) provide the facilities for user-defined filter masks.

Smoothing Filter works basically by taking an average of the intensities of all the pixels in the kernel and assigns this as the new intensity value of the central pixel. The size of the kernel and the mask will determine the degree of smoothing – the larger the kernel, the higher the degree of smoothing introduced, which is associated with a certain degree of blurring (Blur filter). This filter is also an important tool for reducing noise (individual pixels of very high values, distributed randomly across the image area) in an image.

Edge-detection filters are useful in object (shape) recognition, preceding measurements and segmentation. The basic strategy in the operation of these kind of filters is to detect where in the image there are sudden increases or decreases in the intensity levels. These filters can be applied either in one direction (horizontal or vertical) or sequentially in both directions (Sobel filter). The unidirectional filters are equivalent to a unidirectional image shift (with one or several pixels) followed by a subtraction of this image from the original one. In the resulting image, the edges situated 'ahead', in the direction of filter application, will appear brightened, while the trailing edges will be darkened. The Sobel filter performs sequentially a horizontal and then a vertical shift operation, thus outlining the edges in both directions. It is important to note that these filters will be most effective on the edges which are directly perpendicular to the direction of image shift.

3.1.4. Image segmentation

This procedure implies the division of the whole grey level range into a number of discrete sets. The simplest form is the grey level thresholding or binarization, in which a threshold is set and all the pixels with values under the threshold will be displayed as white and all above as black. This procedure is extremely useful in morphological analysis of images, and the binarization is used for selecting the objects of interest – which should have a different intensity level than the background (e.g., in immunostaining, when the labelled structures will have higher intensities) which can then be measured. A more complex form of binarization is called density slicing, in which a lower and an upper threshold can be set, and pixels with values falling in this range will be displayed as black whereas all others will be white.

More complex image segmentation procedures can be applied, which depend on the fact that the image processor boards contain the output **lookup table** (LUT) computational module, which converts each input grey level to a different new one (see above, Contrast Enhancement). With this method, various pseudocolour tables can be applied to an image, to highlight the features of interest (e.g., the standard colour coding for Ca^{2+} measurements with dark blue representing the resting $[Ca^{2+}]_i$ and the hotter red representing the maximal $[Ca^{2+}]_i$).

3.1.5. Arithmetical and logical operations on images

Another set of image operations, especially important for Ca^{2+} measurements, is the arithmetic operations on images. All four major arithmetic operations: addition, subtraction, multiplication and division, can be performed on any set of two or more images, and this amounts to a simple arithmetic of pixel values in the two images. In a similar fashion, one can obtain an average image, starting from a set of images. Another set of image operations is represented by the application of the boolean logical operators AND and OR, or the Min or Max operations (display, for each location, only the minimal (or the maximal) pixel value in the set of image). For Ca^{2+} measurement imaging, the average, ratioing and subtraction are the most relevant and useful operations.

3.1.6. Image masks

During the initial processing of images for later detailed analysis, one useful operation is masking, which involves two steps. In the first step, a set of pixels from a reference image is selected according to a certain criterion to create a mask, and then, in the second step, this mask is applied to the image to be analysed. In this way, the pixels outside the objects of interest are excluded. In a usual implementation, either the brightfield image or, in the example of a fura-2 experiment, the fluorescence image at 380 nm (which has higher intensity) are used to define such a mask.

3.2 Image analysis

The purpose of this stage of experimental procedures is to extract the relevant information. For stand-alone Ca^{2+} measurement experiments, this usually consists in measuring the values of $[Ca^{2+}]_i$ across the time domain. How this is achieved depends on the type of dye used. For ratiometric dyes (or combination of single wavelength dyes co-loaded simultaneously), the process, in principle, is that of simple arithmetic operation of images: ratioing of image A/image B (e.g., for fura-2 measurements, image A is collected at excitation 340nm and image B at excitation 380nm). However, as discussed in the previous chapter, ratioing requires the calibration of the dye, and the ratioing of the images has to make reference to a LUT table, which simply stores a ratio value for each combination of pixel value in image A – pixel value in image B, taking into account the R_{min}, R_{max}, K_d and β values determined during the calibration experiments (see previous chapter for details). Thus, for two individual pixel values in the two respective images, the LUT provides a new 'ratio' pixel value, corresponding to a $[Ca^{2+}]_i$ value.

In the case of non-ratiometric dyes, the standard procedure of image analysis, is to perform an image ratioing, as an actual arithmetic image operation, using the initial image, or an average of them (F_0) obtained in the resting conditions, as denominator for diving images obtained at later times (F_i/F_0 ratio images). While this ratioing operation cannot provide any information about the actual $[Ca^{2+}]_i$ values, it can provide information about the relative (fold-change) $[Ca^{2+}]_i$ changes upon various stimulation protocols in comparison with the resting (reference) condition. This method of ratioing, by removing the effect of different amounts of cellular dye loading for each particular cell under investigation, is a normalization procedure, which allows inter-experimental comparisons. Since in confocal microscopy most of the dyes commonly used are non-ratiometric (e.g., fluo-3), the expression of intracellular Ca^{2+} changes as F_i/F_0 ratios is very commonly encountered. The next step in measuring $[Ca^{2+}]_i$ values on the stack of ratioed images, is to define a number of regions of interest (ROIs) within which the Ca^{2+} values, or the ratio values, are measured and displayed then as a graph having time as the abscissa.

For confocal microscopy, which allows in addition to the scan image, repeated measurements along only one pixel-width line (line-scan mode), another very informative way of presenting the data is the **x-t plot**, in which the scanned line, colour-coded for $[Ca^{2+}]_i$ values, is stacked line after line, along a time ordinate. This type of display is extremely useful in providing immediate information about the initiation points of Ca^{2+} events and also about the rates of Ca^{2+} spread from those initial points.

In addition to these Ca^{2+} measurements-specific methods discussed above, image analysis can investigate a variety of other parameters of the object imagined. A detailed discussion of these methods is beyond the remit of this chapter, but it is important that researchers are aware of them, as multiparameter analysis, linking Ca^{2+} response with a variety of other morphological or immunocytochemical staining parameters, will become, for sure, more and more common. Measurements that can be performed on features of objects (cells) in images can be grouped in five classes:

- brightness
- location
- numbers (counts)
- size, and
- shape.

For each class, a variety of different specific measurements can be made, and there are a number of different ways of performing these operations. Most of the dedicated, and normally quite expensive, image analysis packages (e.g., Image ProPlus, MetaMorph, etc) will offer a wide range of these features. For defining the objects of interest, one can either use manual definition, similar to defining the ROI, as described above, or use automated or semiautomated methods of image processing which will extract, on the basis of a combination of methods presented in a general form, in the previous section, the objects of interest (binarization of image, using thresholds or density slices, edge detection, etc). Identification of cellular structures in an image is not always an easy process, but the advances in computational power render these (semi)automated methods more practical.

The result of these measurements is a numerical output which can then be easily imported into a spreadsheet or a statistical package for extracting the statistical significance. The importance of this approach is in the fact that in this way one can correlate directly, at the level of individual cells, the relationship between Ca^{2+} signalling and cell viability, or cell shape or immunostaining for one protein or another.

References

Brown CW, Shepherd BJ (1995) Graphics File Formats Reference and Guide. Manning (Prentice-Hall), 472 pp.

Glasbey CA, Horgan G (eds) (1995) Image Analysis for the Biological Sciences, John Wiley & Sons, 230 pp.

Russ, JC (1995) The Image Processing Handbook. CRC Press, 674 pp.

Aequorin Measurements of Cytosolic Ca^{2+}

C. JANE DIXON and ANNE K. GREEN

Introduction

Aequorin is a 21 kDa luminescent protein, present in photocytes of the jellyfish Aequorea forskalea, which generates blue-green light upon binding Ca^{2+} ions. Aequorin can be used to measure cytosolic free Ca^{2+} ($[Ca^{2+}]_i$) in single cells by injecting the purified protein, and then measuring the light emitted from the cell. This protocol describes the transfer of a single cell into an optically-flat capillary (microslide), microinjection of aequorin into the cell, and detection of the signal. Inside the microslide, the cell is suspended within an agarose gel. Such an arrangement enables access to the cell for microinjection, and facilitates transfer of the cell between apparatus throughout the experimental procedure. During recording of the signal from the aequorin-injected cell, culture medium containing agonists etc. is superfused over the microslide (see subprotocol 6). Inside the agarose gel, the cell is protected from the flow of medium, but since the diffusion path through the agarose to the cell is only ~50 μm, the cell is readily accessed by the superfused culture medium and agonists.

Subprotocol 5 describes the microinjection of aequorin into single cells set up in microslides. An unusual feature of this technique is that the cell is injected from the side rather than from above, thus minimising mechanical damage to the cell. This ap-

C. Jane Dixon, The University of Liverpool, New Medical School, Department of Human Anatomy and Cell Biology, Ashton Street, Liverpool, L69 3GE, UK
Anne K. Green, The University of Liverpool, Department of Human Anatomy and Cell Biology, Ashton Street, Liverpool, L69 3GE, UK
(*phone* +44-151-794-5468; *fax* +44-151-794-5440;
e-mail annekg@liverpool.ac.uk)

proach is particularly advantageous for the microinjection of cardiomyocytes, which are especially sensitive to mechanical damage. During the microinjection procedure, the operator is able to manually control the extent and speed of cell impalement, and the volume and speed of the microinjection. The procedure can thus be optimised for each individual cell. The microinjection apparatus is not commercially available but is custom-built. Detailed instructions for the construction of this apparatus have been published previously (Cobbold and Lee, 1991). This subprotocol includes a description of the microinjection apparatus and instructions for its use. Practical hints and suggestions are included, which may also be of use to workers using commercially available apparatus.

The microslide containing the aequorin-injected cell is transferred into a stainless steel perfusion chamber, held immediately in front of a low noise photomultiplier. Photon counts are collected, normalised retrospectively by computer, and converted to $[Ca^{2+}]_i$. The apparatus for signal detection is not commercially available. Instructions for its construction and use are given in subprotocol 6.

Subprotocol 1
Preparing Cells for Microinjection

Materials

Equipment

- 2 stereomicroscopes (magnification x32 to x80), with stages maintained at 37°C
- Block heater (Grant Instruments)

Supplies

Preparation of microslides: Optically flat glass capillaries (microslides; Camlab Ltd.), with a path length (floor to ceiling) of 0.1 mm are cut into ~5 mm lengths (see Figure 1a) and washed by immersion in conc. nitric acid for at least one hour. Care must be taken to ensure that each microslide fills with nitric acid. The microslides are rinsed thoroughly with distilled water and then stored in absolute ethanol until required. The washed microslides may be stored in ethanol for several days.

Preparation of soda glass micropipettes: Soda glass micropipettes are prepared, using a vertical microelectrode puller (Harvard Apparatus Ltd.), from soda glass melting point tubes (length 100 mm, diameter 1.8 - 2.0 mm; Samco; Fisher Scientific Ltd.).

- Liquid paraffin (SpectrosoL grade; Merck Ltd.)
- Polystyrene tissue culture dishes (60 mm diameter x 15 mm).

Solutions

Preparation of agarose-containing media: To prepare the agarose solution for suspending cells in agarose gel in microslides (see below), 1.2% (w/v) FMC SeaPlaque agarose (Flowgen Instruments Ltd.) is dissolved in the culture medium. SeaPlaque agarose gels below 30 °C, and, once gelled, will not remelt at 37 °C for several hours. During "setting-up", this agarose solution is kept at 37 °C in a block heater, with an overlay of SpectrosoL liquid paraffin (to prevent evaporation and loss of CO_2 from the medium).

To increase the viscosity of the medium in the stock dish of cells, FMC SeaPrep agarose (1.7%) is added. SeaPrep agarose will not gel above 17 °C.

To avoid boiling heat-susceptible media, agarose powder is dissolved, at double concentration, by boiling in distilled water. Double-strength medium is then added upon cooling to 37 °C. The composition of the culture medium is dictated by the particular cell type. For example, for rat hepatocytes we use Williams' Medium E. BSA (fatty acid adsorbed; 0.05 - 0.25%) is added to the culture medium to aid membrane resealing following microinjection. For each cell type the optimum concentration of BSA must be determined; too much BSA can prevent penetration of the plasma membrane by the micropipette.

Procedure

Preparation of a stock dish of single cells

A 60 mm diameter plastic tissue culture dish is coated with a thin underlay of 1.2% FMC SeaPlaque agarose in culture medium, which is then gelled by cooling to 4 °C. This agarose underlay prevents the cells from adhering to the dish. Using a pasteur pipette, 1.2% SeaPlaque agarose is expelled to form, upon gelling, a 30 mm diameter well in the centre of the dish. Approximately 1 ml of a dilute suspension of single isolated cells (~10^3 cells ml^{-1}) in culture medium containing 1.7% SeaPrep agarose (to add viscosity to the medium) is added to the well. This arrangement retains the cell suspension in the centre of the dish and thus permits access to the cells with a micropipette held at a shallow angle. The dish is immediately overlaid with approximately 3 mm depth of SpectrosoL liquid paraffin (to prevent evaporation and loss of CO_2) and placed in a CO_2 incubator at 37 °C.

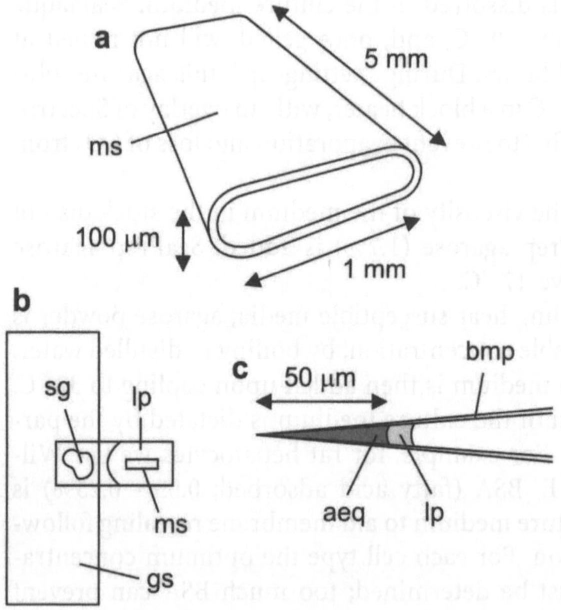

Fig. 1. Preparing cells and filling pipettes for microinjection (a) Dimensions of a glass microslide (ms) for setting up a single cell for microinjection. (b) A rectangular piece of glass coverslip is affixed to a glass microscope slide (gs) by silicone grease (sg). The microslide is attached and then filled with medium containing agarose and covered with liquid paraffin (lp). A single cell is transferred into the filled microslide. (c) Borosilicate micropipettes (bmp) are filled by dipping into a droplet of dialysed aequorin. A short column of liquid paraffin (lp) enters the tip followed by the aequorin (aeq).

Transferring a single cell into a microslide

1. A rectangular piece of coverslip (22 x 10 mm) is affixed to the edge of a glass microscope slide with silicone grease, with the free end of the coverslip cantilevered out and tilted slightly downwards.

2. A flame-dried microslide is transferred onto the coverslip, with the opening of the microslide ~2 mm from the edge of the free end of the coverslip (see Figure 1b).

3. The microslide is viewed through a stereomicroscope. Using a blunted soda glass micropipette with a tip bore of approx. 50 μm, a small amount of medium containing 1.2% SeaPlaque agarose is applied close to the edge of the microslide at the opposite end to which the cell is to be inserted. The agarose flows by capillarity between the microslide and the coverslip and, once gelled, serves to anchor the microslide in position. Care should be taken to avoid the insertion of this agarose anchor directly underneath the region of the microslide into which the cell is to be introduced, since gelled agarose on the underside of the microslide will prevent the clear focussing on the cell which is essential for microinjection.

4. The microslide is then filled with this agarose solution, leaving a small meniscus. Once filled, the microslide is immediately covered with liquid paraffin to prevent evaporation. The microscope stage is maintained at 37 °C to prevent premature gelling of the agarose.

5. The stock dish of cells is viewed through an adjacent stereomicroscope, the stage of which is maintained at 37 °C. A cell of healthy appearance is selected and sucked into the blunted soda glass pipette containing SeaPlaque agarose: Suction is applied to the pipette to achieve a flow of medium containing the cell into the pipette. Immediately the cell enters, the pipette is rapidly withdrawn from the dish and suction is released; this prevents entry of an excessive volume of SeaPrep agarose-containing medium into the pipette. Thus, the cell comes to rest within 0.1-0.2 mm of the tip of the pipette.

6. The cell is then immediately expelled into the opening of the microslide. The cell should be positioned within ~50 μm of the meniscus of the SeaPlaque agarose-containing medium. Rod-shaped cardiomyocytes should be aligned so that their length is parallel with the aperture of the microslide.
7. The microslide containing the cell is then refrigerated at 4 °C for 3 minutes to gel the agarose.
8. It is then placed in an incubator at 37 °C until required for microinjection.

Subprotocol 2
Preparation of Aequorin Stock Solution

Materials

Equipment

- Stereomicroscope (x10 to x45)
- Vertical microelectrode puller (Harvard Apparatus Ltd.)

Supplies

- Native aequorin of quality can be purchased from Prolume Ltd and Research Biochemicals International (available from Sigma-Aldrich Company Ltd). Recombinant aequorin is available from Molecular Probes.
- Borosilicate glass micropipettes are pulled from borosilicate glass capillaries (length 100 mm; bore 1.16mm; wall thickness, 0.42mm; Clark Electromedical Instruments).
- Liquid paraffin (SpectrosoL grade; Merck Ltd.)

Solutions

Preparation of solution for dissolving aequorin: Aequorin is dissolved in 'aequorin dialysis buffer' (ADB). Ca^{2+} levels of this buffer must obviously be kept low to avoid discharge of aequorin; ultra-pure distilled water and EDTA- and EGTA-buffering are used. ADB is prepared and stored in a 500 ml tissue culture flask; glassware is avoided during preparation. The buffer composition is as follows: KCl, 150 mM; PIPES, 1 mM; EDTA, 100 μM; EGTA, 25μM. The solution pH is adjusted to 7.2 by removing aliquots rather than directly introducing the pH electrode, as aequorin is catalytically destroyed by traces of silver and mercury. ADB, stored at 4 °C, can be used over several months. Adequate pH buffer must be included in the ADB to avoid a rise in pH consequent upon exchange of H^+ for Na^+ in the borosilicate glass pipettes used for microinjection (see subprotocol 4); this can lead to a very rapid discharge of aequorin in the pipette and a toxic pH in the injected solution.

Procedure

1. 1 mg freeze-dried aequorin is dissolved in 15 µl ADB and the solution immediately covered with liquid paraffin.
2. The aequorin solution is sucked into an externally-silanized (Repelcote) borosilicate glass micropipette whose end has been snapped off to give a ~0.1 mm bore. This procedure is carried out under the stereomicroscope (~20 x). Movement into and out of the pipette is finely controlled by a rubber bulb of ~5 cm diameter which is connected to the pipette by silicon rubber tubing. The apex of the bulb is pierced with a small hole which can be occluded by the forefinger to control the transmission of pressure or suction to the pipette. The aequorin stock solution is divided into ~15 aliquots as droplets formed on the bases of 60 mm diameter plastic tissue culture dishes containing about 5 mm depth of liquid paraffin. To aid attachment of the droplet, the plastic can be lightly scratched with the pipette tip and pre-wetted with ADB.
3. These droplets of stock aequorin solution are stored at -70 °C and are stable for many months. Aliquoting the aequorin amongst several Petri dishes reduces the number of freeze-thaw cycles each drop is subjected to. When a dish is removed from the -70 °C freezer, it is placed in a dessicator containing silica gel to prevent formation of droplets of condensation on the oil surface.

Subprotocol 3
Aequorin Microdialysis

Materials

Equipment

Microdialysis apparatus: The stock solution of aequorin cannot be injected into cells because the concentrations of EDTA and KCl are too high. Aliquots are dialysed on a micro-scale (~200 nl) against ADB. The microdialysis apparatus (see Figure 2a) is constructed using Teflon-coated platinum wire (130 µm diameter; PT-5T, Clark Electromedical Instruments). The ends of two equal lengths (~2 cm) of wire are shaped to form v-shaped clefts and the wires glued at their opposite ends to a piece of plastic cut from the base of a Petri dish and holds the tubule (0.5 cm above the plastic base. If the plastic base is cut from the edge of a Petri dish the wall can be gripped by forceps, facilitating transfer of the apparatus between solutions.

Supplies

The microdialysis tubule we routinely use is no longer available (Biofiber 50A; Bio-Rad). A suitable alternative is 'Microtubing' supplied by Medicell International (London). This tubing is transparent with a bore ~180 µm and a cut-off of ~5kd.

Solutions

- Just prior to dialysis, glutathione (GSH) is added to a 20 ml aliquot of ADB (see subprotocol 2) to give a final concentration of 1 mm. This improves the flow of aequorin into and out of the pipette during microinjection.
- EDTA-PIPES (1 mm each; pH 7.0).

Procedure

1. A ~1 cm length of tubule is held in the v-shaped wire clefts. The tubule is clamped lightly in place in the microdialysis apparatus using fine forceps to compress the wire whilst ensuring the bore is not occluded. The microdialysis apparatus is immersed for several minutes in EDTA-PIPES in a deep, 50 mm diameter plastic tissue culture dish.

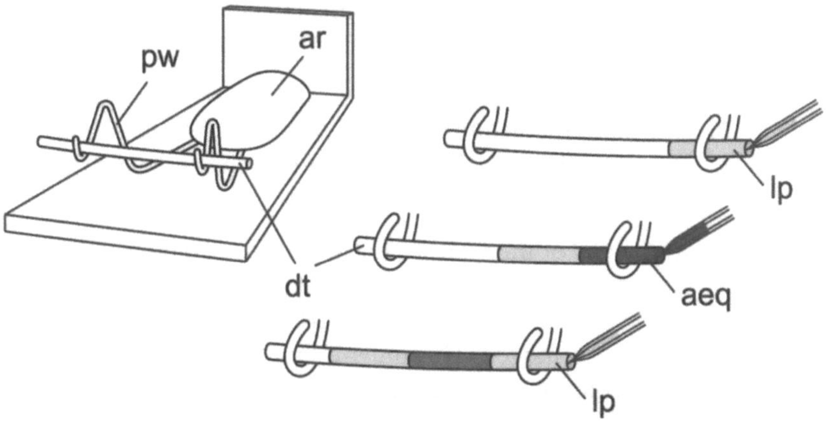

Fig. 2. Microdialysis of aequorin The stock solution of aequorin must be dialysed to reduce KCl and EDTA levels before microinjection. (**a**) The micro-dialysis apparatus is constructed from teflon-coated platinum wire (pw) attached to a section of plastic petri dish by araldite (ar). (**b**) A piece of dialysis tubing (dt) is held in the v-shaped clefts and is filled with a column of liquid paraffin (lp), followed by aequorin (aeq) and then a second column of liquid paraffin, before immersion in aequorin dialysis buffer.

2. The dialysis tubule in its wire cradle is then placed under the stereomicroscope (~20 x). The EDTA-PIPES buffer in the tubule lumen is removed using a blunted borosilicate glass micropipette. The tubule is then washed several times with ADB containing GSH, by filling and emptying the lumen.
3. A column of liquid paraffin is introduced into the dialysis tubule to ~1/3 of its length. A fresh borosilicate micropipette with the very end of the tip snapped off to give a bore of ~50 µm is loaded with liquid paraffin followed by ~200 nl of stock aequorin solution. The volume is judged by the length of the column of aequorin in the pipette tip. The aequorin solution is expelled into the lumen of the microdialysis tubule and is followed by a second column of liquid paraffin.
4. The dialysis apparatus is then submerged in about 1 cm depth of ADB containing GSH in a deep, 50 mm diameter plastic tissue culture dish. Because of the high osmotic pressure there is a tendency for the aequorin solution to swell within the first seconds of dialysis, with the risk of displacing the paraffin plugs and loss of the aequorin. If this starts to

happen the tubule should immediately be removed from the ADB and the column of aequorin allowed to evaporate back to its original size.

5. Dialysing the aequorin for 5-10 minutes is sufficient to reduce the KCl and EDTA concentrations to levels that can be tolerated by cells.

6. At the end of the dialysis, the apparatus is removed from the ADB and under the stereomicroscope the aequorin is sucked out of the tubule lumen with a fresh pipette whose very end has been removed. This pipette should first be loaded with a short column of liquid paraffin to give greater control over suction into the pipette. The process can be facilitated by cutting the tubule close to the column of aequorin whilst still immersed, so that the aequorin solution is more accessible.

7. The aequorin solution is placed as a droplet under ~5 mm liquid paraffin in a 60 mm diameter plastic tissue culture dish. If a sufficient volume of aequorin has been dialysed it can be subdivided into separate dishes. The droplets are stored at -20 °C and used for several days' experiments, although a gradual loss of activity occurs.

Subprotocol 4
Filling Micropipettes for Injection

Materials

Equipment Vertical microelectrode puller (Harvard Apparatus Ltd.)

Supplies Micropipettes are pulled from borosilicate glass capillaries (see subprotocol 2). The capillaries are not washed, but are fitted into the puller using forceps to avoid contaminating the region which will form the tip. The settings used produce pipettes with relatively long tapering tips (~11 mm from tip to full diameter). The bore of the pipette tip has been measured under SEM as ~0.3 µm. Pipettes are used straight from the puller without washing.

Procedure

Pipettes are filled by simply dipping the tip for a few seconds into the droplet of dialysed aequorin viewed under the stereomicroscope (45 x). As the pipette is lowered through the liquid paraffin a short column enters the pipette and is followed by a volume of aequorin (Figure 1c) that is sufficient for a single injection.

Subprotocol 5
Microinjection of Aequorin Into a Single Cell

Materials

Equipment

Microinjection apparatus: An inverted microscope (Nikon Diaphot) with Nomarski optics is used. A 40 x LWD objective and 15 x eyepieces give an adequate image size for microinjection of a single cell. Attached to the microscope stage are mechanical fittings which are used to control the position of a thin brass platform, upon which the microslide is placed, mounted on the end of a piezoelectric bender element ('Biomorph' PZT-5N, Morgan Matroc Ltd.). The piezoelectric element is controlled by a voltage-reversing switch via a variable voltage supply. This arrangement allows the piezoelectric motion of the bender element to move the brass platform, thus rapidly throwing the cell onto the pipette tip. The variable voltage supply allows adjustment of the distance of movement of the brass platform and hence the extent of impalement of the cell. For hepatocytes this is set to provide a movement of ~3 µm. A second switch provides a voltage of opposite polarity; the ensuing movement of ~15µm in the reverse direction enables the cell to be rapidly and cleanly jerked away from the micropipette upon completion of the microinjection.

Movement of the micropipette is achieved by a mechanical manipulator (Leica UK Ltd.), which is mounted to the right of the microscope; this can be controlled by X, Y and Z controls and also by a joystick. A brass pipette holder is clamped horizontally into the micromanipulator. The pipette holder is connected to a supply of nitrogen at 4 atm. A rubber seal (Camlab Ltd.) inside the base of the pipette holder ensures that the gas pressure is directed through the micropipette. The pressure reaching the pipette is directed through one of two regulators by operating a solenoid valve through a foot switch.

Before injection, with the valve in a closed position, a pressure of ~0.5 atm is transmitted to the pipette. After impaling the cell (see microinjection procedure), depression of the foot-switch diverts pressure through the second, lever-controlled, regulator. The pressure is regulated by the movement of the lever with the left hand. The lever is thus raised until sufficient pressure is applied to expel the aequorin from the pipette.

Procedure

1. A film of silicone grease attaches the coverslip holding the microslide, covered with a drop of liquid paraffin, to the brass platform on the microscope stage.
2. The microslide is aligned in the left-hand side of the field of view, focussing, with the 20 x objective, on the outer edge of the glass wall of the microslide.
3. The aequorin-filled micropipette (see subprotocol 4) is placed in the brass holder, mounted on the micromanipulator.
4. Using the X, Y and Z controls, the micropipette is positioned alongside the edge of the microslide. Using the 20 x objective, the vertical position of the micropipette is adjusted so that its tip is parfocal with the edge of the microslide. In this vertical position the tip will be midway between the floor and ceiling of the microslide.
5. Without adjusting its vertical position, the micropipette is repositioned close to the aperture of the microslide so that it is pointing at the cell.
6. Regulator R1 is adjusted to apply a gas pressure of 0.5 atm pressure to the micropipette. This prevents entry of Ca^{2+}-containing medium from the gelled agarose entering the tip of the pipette; this would discharge the aequorin.

7. Using the X, Y controls, the micropipette is advanced into the microslide until it is within ~30 μm of the equator of the cell.

8. Using the 40 x objective, fine adjustment of the vertical position of the micropipette tip is performed so that its is parfocal with the cell equator.

9. The micropipette is then slowly advanced, using the joystick, until its tip dimples the cell surface by ~1-2 μm. Cell penetration is achieved using the piezoelectric bender (as described above) to rapidly throw the cell onto the micropipette tip.

10. The foot switch is depressed and the lever is slowly raised until the pressure is sufficient to expel the aequorin into the cell. A swirl of cytoplasm near to the pipette tip indicates that the aequorin has entered the cell. The injection is terminated by pressing the switch which applies a voltage, in the reverse direction, to the piezoelectric bender. This rapidly throws the cell backwards, away from the pipette tip. The operator can therefore limit the volume of the injection, by removing the cell from the pipette before all of the aequorin has been expelled. A rough estimation of the injection volume can be made by estimating the size of the cell, and using an eyepiece graticule to estimate the volume of aequorin in the pipette. We aim to inject less than 1% of the cell volume. The micropipette is then withdrawn from the microslide using the X, Y controls.

11. The cell is examined for any signs of injury. Typical signs of injection damage to hepatocytes are membrane blebbing or leakage of cytoplasm. Damaged heart cells often lose their rod-shape. Damaged cells are discarded; cells showing no visible signs of injury are returned to the incubator for ~10 minutes, then re-examined before they are transferred to the apparatus for signal detection.

Subprotocol 6
Signal Detection From a Single Aequorin-Injected Cell

Materials

Equipment

1. **Perfusion chamber (Figure 3):** The chamber into which the microslide will be placed is made by machining a polished cavity of approximate dimensions 8 mm length x 4 mm width x 2 mm depth in a disc of 316 stainless steel. The

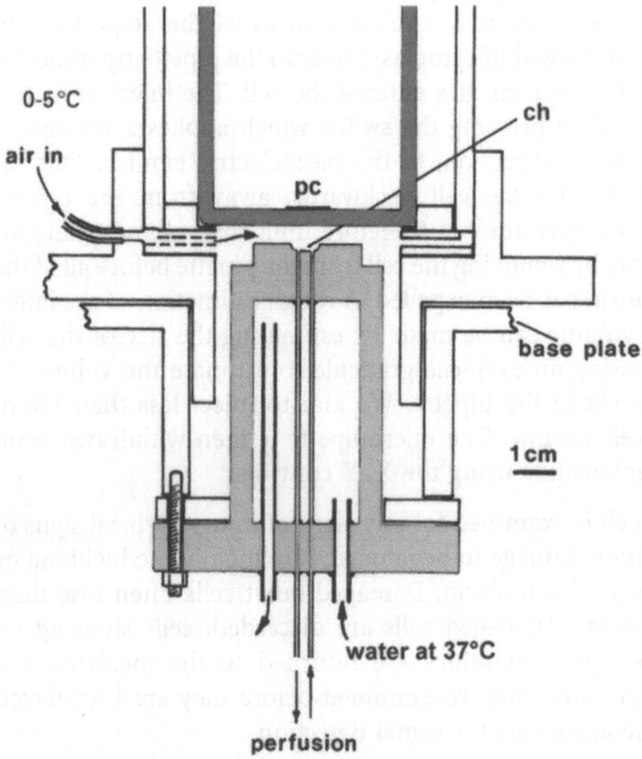

Fig. 3. Diagrammatic representation of perfusion chamber and photomultiplier housing. The microslide containing the injected cell is placed in a perfusable chamber (ch) which is positioned under the photocathode (pc) of the photomultiplier which is cooled to 0-5 °C. The chamber is held at 37 °C by a water jacket. A flow of cold air prevents the photomultiplier from being warmed.

use of a polished steel rather than a transparent perspex chamber minimises background light. Small bore (~1 mm) stainless steel tubes at each end of the chamber allow perfusion of medium through the chamber at rate of ~100 µl min^{-1}; this is controlled by suction from a variable-flow peristaltic pump. The plumbing allows media to be changed by syringe suction from a flask to a "T" piece, mounted below the chamber's water jacket. The water jacket is maintained at 37 °C by a thermocirculator. The chamber is sealed with a 22 mm diameter, thickness no. 1, glass coverslip (dark-adapted) on a film of liquid paraffin (dark-adapted).

2. **Photomultiplier housing (Figure 3):** The perfusion chamber is positioned directly below the photocathode of the photomultiplier, leaving ~3 mm air space to prevent disturbance of the electric field of the tube by the stainless steel chamber. Shims can be used to adjust this separation if necessary. The photomultiplier is cooled to ~0 °C. This is achieved by installing the perfusion chamber, photomultiplier and housing inside a constant-temperature incubator. We have found that a domestic "larder" refrigerator can be adapted for this purpose and is an economical and reliable alternative. A tendency for the photocathode to be heated by the warm-water jacket of the perfusion chamber is prevented by a flow of cold air across the face of the photomultiplier and down the sides of the water jacket. This is provided by an aerator pump.
   The chamber can be lowered from its position below the photomultiplier (e.g. to insert the microslide). While this is being done, a rubber bung is used to block the opening to the photomultiplier. The photomultiplier is never exposed to light even when switched off; this minimises phosphorescence.

3. **Photomultiplier:** The photomultiplier (Electron Tubes; type 9789B) has a 10 mm diameter bi-alkali photocathode. Tubes are requested from the manufacturer with a low dark count rate, ideally between 1 and 4 counts per second (cps). Cooling the photomultiplier to ~0 °C has no effect on its quantum efficiency, but does reduce the thermal generation of electrons. However, cooling below 0 °C leads to an increased

resistance of the photocathode, and counting becomes erratic. The photomultiplier housing is type B2 from Electron Tubes.

4. **High voltage supply:** The HT supply (Electron Tubes; model PM28B) provides regulated voltage for the tube, which is set to a voltage which, from the manufacturer's data, gives a sensitivity of 5000 A lumen $^{-1}$, usually around 1200 to 1500 V for a 9789 tube. The HT must be switched off whenever the incubator door is opened.

5. **5. Amplifier-discriminator and photon-counting board:** A photomultiplier needs only two wires. One supplies the high voltage; the other (the anode) carries the current generated as electrons flow up the dynode chain. One electron released from the photocathode by a photon generates a burst of, typically, 10^8 electrons, lasting typically ~5 ns. An amplifier-discriminator (Electron Tubes; type AD1), positioned adjacent to the photomultiplier housing amplifies the current pulses, and discriminates between slow fluctuations in current and the single photon-induced pulses. These pulses are converted into standard pulses which can be counted by the photon counting board (Electron Tubes; type CT1) fitted into the PC. Software for displaying the data (counts per second) on screen was written by Dr. K.S.R. Cuthbertson and is available from the authors.

6. **Chart recorder:** A digital to analogue converter provides a chart recorder output directly from the computer for on-line visualization of the cell's response. Time constants for exponential smoothing of the data are under software control and can be altered between 1 and 100 sec.

Procedure

1. The microslide containing the aequorin-injected cell is removed from the coverslip, using a pair of forceps, and rinsed in culture medium to remove the liquid paraffin.

2. The microslide is transferred to the cavity of the perfusion chamber. The chamber is sealed with a circular coverslip over a film of liquid paraffin.

3. The perfusion chamber is positioned immediately below the photomultiplier.

4. The photomultiplier is supplied with high voltage, and the current generated by aequorin luminescence is passed through the amplifier-discriminator.

5. The brief (~5 ns) pulses of current generated, each of which corresponds to a single photon of aequorin luminescence are counted by computer fitted with the photon counting board. Photon counts are deposited in 50 ms bins ("fast data"). A "slow data" file is also generated by integrating signals over 1 second intervals.

6. A resting signal is recorded when the cell is superfused with culture medium alone. At resting Ca^{2+} levels of ~200 nM about 10^{-6} of the total injected aequorin will be consumed per second, generating resting signals of ~0.3 - 2.0 cps which is superimposed on a background of about 1 to 4 cps. This resting signal is detectable over ~5 minutes plotting cps smoothed exponentially with a long time constant (30 s).

7. During the experiment, agonists etc are applied through exchange of medium. This is achieved swiftly using gentle syringe suction to bring the medium containing the substance of interest to the T-piece just beneath the chamber as described above. The time constant is lowered to 1 s so that rapid changes in $[Ca^{2+}]_i$ are captured on the chart recording.

8. The amount of aequorin injected is not standardized. Instead, normalization of the signals requires the total aequorin content of the cell to be determined at the end of each experiment. This is achieved by discharging the aequorin by lysing the cell with distilled water whilst the photomultiplier is still counting. The total number of photon counts recorded from a single hepatocyte is typically about 300 000 whilst that from a cardiomyocyte can be 2 000 000.

9. Signals are normalized retrospectively by computer, by calculating cps as a fraction of the total counts remaining after the subtraction of background. This computed fractional rate of consumption of aequorin can then be expressed as Ca^{2+} using *in vitro* calibration data.

Subprotocol 7
Data Presentation

Procedure

Normalized signals are calibrated in terms of Ca^{2+} using data on the rate of consumption of aequorin at 37 °C using Ca^{2+}-EGTA buffers mimicking the intracellular milieu, with $[Ca^{2+}]$ ranging

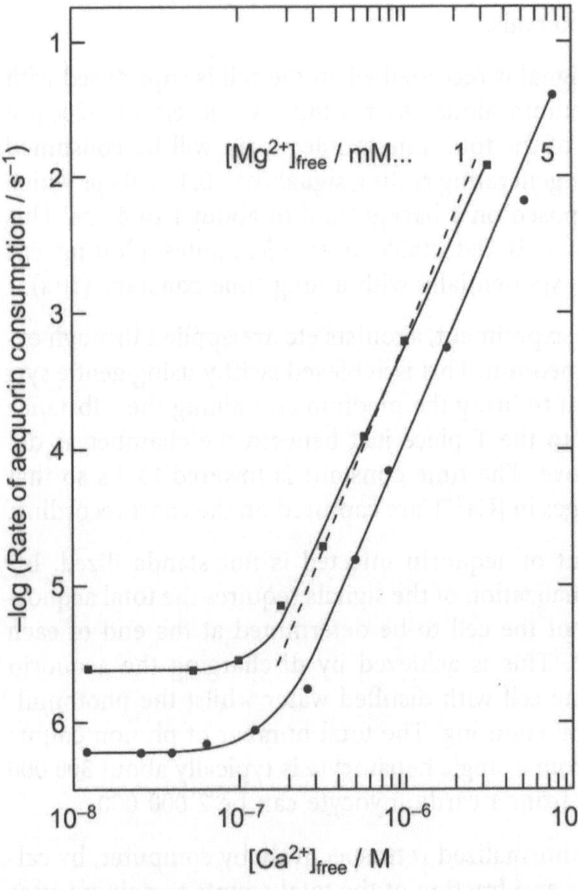

Fig. 4. Calibration data for aequorin luminescence. The relationship between the rate of aequorin consumption and $[Ca^{2+}]$ *in vitro* was determined in 2mM EGTA buffers containing 150 mM KCl, 10 mM PIPES, and $[Mg^{2+}]$ as indicated, at 37 °C. Reproduced with permission from Cobbold, P.H. & Rink, T.J. (1987) Biochem. J. 248, 313-328. Biochemical Society.

between 10^{-8} M and 10^{-5} M. The apparatus required, and the protocol used, for collection of calibration data have been described previously (Cobbold and Lee, 1991). These calibration data (Figure 4) have been incorporated into the data-analysis software, enabling the fractional rate of consumption of aequorin to be plotted as $[Ca^{2+}]_i$. Copies of this software, written by Dr. K.S.R. Cuthbertson, are available from the authors. Data are plotted with exponential smoothing. The time constant applied can be varied to allow faithful representation of rapid changes in $[Ca^{2+}]_i$ without excessive noise in the interim periods when Ca^{2+} is at the resting level. Thus resting signals are plotted with a long time constant (typically 10-15 s, for "slow data"), which switches to a much shorter value (typically 1s for "slow data") when the signal exceeds a pre-determined value (Figure 5). Once the signal falls below a second threshold value the longer time constant is re-applied. Detailed plots, for example to follow the time-course of individual $[Ca^{2+}]_i$ oscillations, can be produced using the fast data file (Figure 6).

Results

Although much less widely used, aequorin has a number of advantages over fluorescent indicators for the measurement of $[Ca^{2+}]_i$. The steep relationship between aequorin luminescence and Ca^{2+} concentration makes aequorin ideal for recording the dynamic changes in $[Ca^{2+}]_i$ in the physiological range that occur during $[Ca^{2+}]_i$ oscillations. Aequorin however lacks precision for monitoring small changes around the resting level. Because the luminescent reaction is "once-only" in nature the duration over which $[Ca^{2+}]_i$ at high concentrations can be measured is limited. However, where transient rises in $[Ca^{2+}]_i$ are being monitored, exhaustion of aequorin is not a limiting factor. Furthermore, the use of aequorin rather than a fluorescent probe avoids the problem of buffering of Ca^{2+} by the probe itself. The need to microinject aequorin into cells need not be a limit on productivity. Microinjection of a cell takes less than 5 min, and the success rate for hepatocytes and cardiomyocytes is very high. Once microinjected, aequorin remains in the cytosol and problems of intracellular compartmentalisation and dye leakage associated with some fluorescent probes are not there-

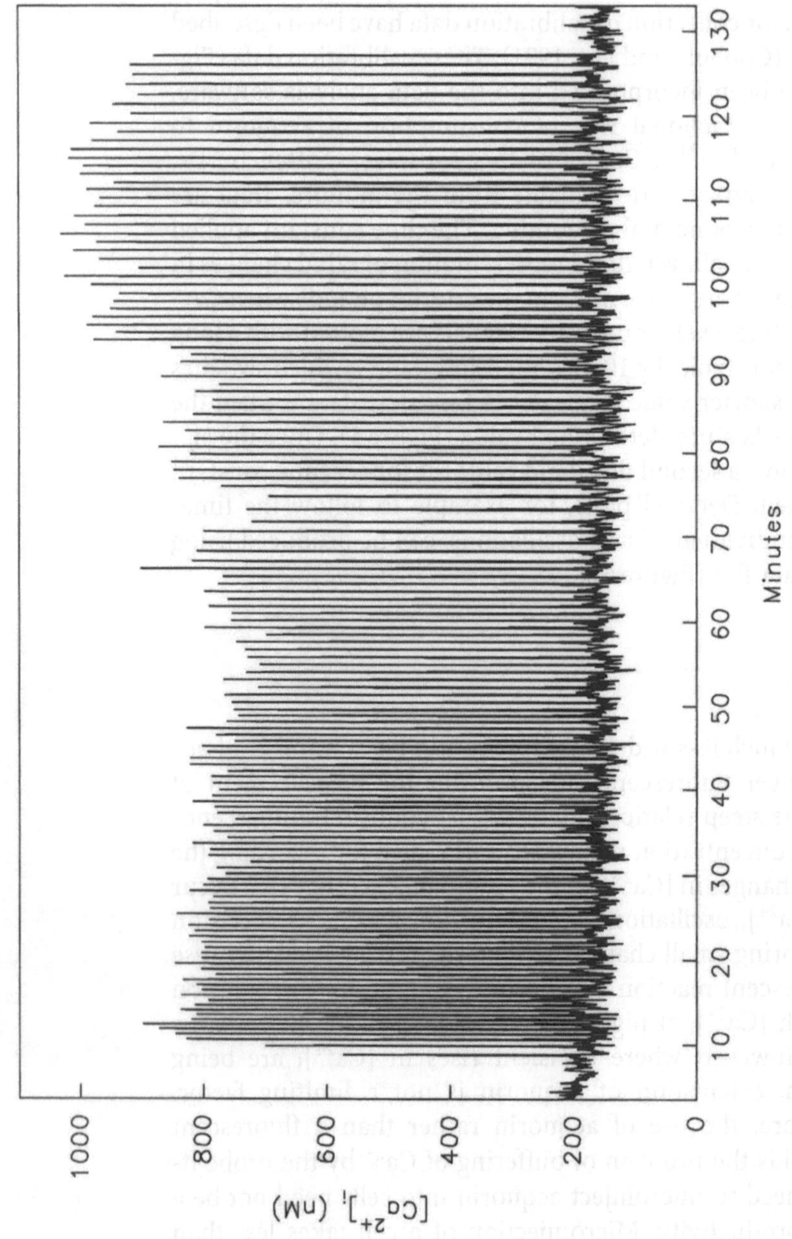

Fig. 5. $[Ca^{2+}]_i$ transients recorded from a single aequorin-injected hepatocyte. The cell was subjected to various experimental protocols which affected the frequency and amplitude of $[Ca^{2+}]_i$ transients. The "slow data" file was plotted using time constants of 12s for resting levels, and 1s for transients. The cell continued to produce oscillations for a further 3 hours.

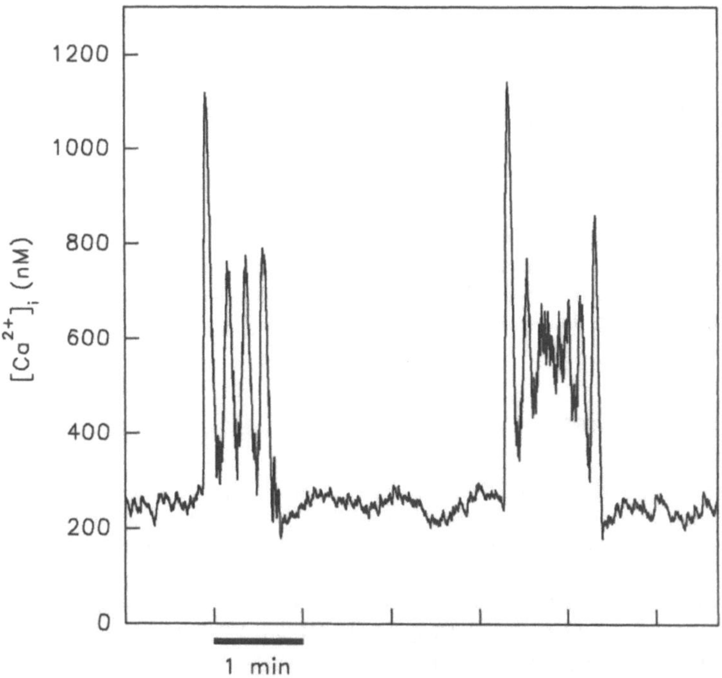

Fig. 6. Detailed plot of complex $[Ca^{2+}]_i$ transients induced by UTP in a single aequorin-injected hepatocyte. The "fast data" file was analysed and data points at 100 ms intervals plotted using time constants of 5 s for resting levels, and 0.6 s for transients.

fore encountered. In addition photo-bleaching of fluorescent probes imposes further constraints on experimental duration. As illustrated in Figure 5, oscillations can be recorded from a single aequorin-injected cell for a period in excess of 5 hours. Data is collected at 50 ms intervals over the entire recording period. This allows the time-course of complex changes in $[Ca^{2+}]_i$ to be followed (Figure 6).

The signal detection instrumentation is much simpler and cheaper than that required for fluorescence measurements of Ca^{2+}, and has been constructed and successfully used in several laboratories. The design of the perfusable chamber can be adapted to hold coverslips, enabling recordings from cells transfected with targeted aequorin (e.g. Rizzuto et al. 1993, Lawrie et al. 1996, Kendall et al., 1996). The apparatus has also been used to monitor Ca^{2+} efflux from hepatocytes using aequorin as the extracellular Ca^{2+} reporter (Green et al., 1997), to monitor ATP release from cells (Dixon et al., 1999), and to measure cytosolic ATP levels in single hepatocytes (Koop and Cobbold, 1993) and cardiomyocytes (Allue et al., 1996), using luciferase as the ATP reporter.

Acknowledgements. We are grateful to The Wellcome Trust for funding. Our thanks to Mr. R. Read for help with the illustrations, and to Professor P. H. Cobbold who developed this technique. We are also indebted to the late Dr. K. S. R. Cuthbertson for writing the software.

References

Allue I, Gandelman O, Dementieva E, Ugarova N, Cobbold P (1996) Evidence for rapid consumption of millimolar concentrations of cytoplasmic ATP during rigor-contracture of metabolically compromised single cardiomyocytes. Biochem. J. 319: 463-469.

Cobbold PH, Lee JAC. (1991) Aequorin measurements of cytoplasmic free calcium. In: McCormack JG, Cobbold PH (eds.) Cellular calcium: a practical approach. Oxford: IRL Press, pp 55-81.

Dixon CJ, Bowler WB, Littlewood-Evans A, Dillon JP, Bilbe G, Sharpe GR, Gallagher JA (1999) Regulation of epidermal homeostasis through $P2Y_2$ receptors. Br. J. Pharmacol. 127: 1680-1686.

Green AK, Cobbold PH, Dixon CJ (1997) Effects on the hepatocyte $[Ca^{2+}]_i$ oscillator of inhibition of the plasma membrane Ca^{2+} pump by carboxyeosin or glucagon-(19-29). Cell Calcium 22: 99-109.

Kendall JM, Badminton MN, Sala-Newby GB, Wilkinson GWG, Campbell AK (1996) Agonist-stimulated free calcium in subcellular compartments. Delivery of recombinant aequorin to organelles using a replication deficient adenovirus vector. Cell Calcium 19: 133-142.

Koop A, Cobbold PH (1993) Continuous bioluminescent monitoring of cytoplasmic ATP in isolated rat hepatocytes during metabolic poisoning. Biochem. J. 295, 165-170.

Lawrie AM, Rizzuto, Pozzan T, Simpson AWM (1996) A role for calcium influx in the regulation of mitochondrial calcium in endothelial cells. J Biol Chem 271: 10753-10759.

Rizzuto R, Brini M, Murgia M, Pozzan T (1993) Microdomains with high Ca^{2+} close to IP_3-sensitive channels that are sensed by neighbouring mitochondria. Science 263: 744-747.

Suppliers

Camlab Ltd.
Nuffield Road
Cambridge CB4 1TH
U.K.
Tel: 01223 424222
Fax: 01223 420856
Email: info@camlab.co.uk

Clark Electromedical Instruments
PO Box 8
Pangbourne
Reading RG8 7HU
U.K.
Tel: 0118 984 3888
Fax: 0118 984 5374

Electron Tubes Ltd.
Bury St.
Ruislip
Middlesex HA4 7TA
U.K.
Tel: 01895 630771
Fax: 01895 635953
Email: pmt@electron-tubes.co.uk

Fisher Scientific Ltd.
Bishop Meadow Road
Loughborough
Leics. LE11 0RG
U.K.
Tel. 01509 231166
Fax: 01509 231893

Flowgen Instruments Ltd.
Broad Oak Enterprise Village
Broad Oak Road
Sittingbourne
Kent ME9 8AQ
U.K.
Tel: 01795 429737
Fax: 01795 471185

Grant Instruments (Cambridge) Ltd.
Shepreth
Cambridgeshire SG8 6GB
U.K.
Tel: 01763 260811
Fax: 01763 262410
Email: sales@grantinst.co.uk

Harvard Apparatus Ltd.
Fircroft Way
Edenbridge
Kent
U.K.
Tel: 01732 864001
Fax: 01732 863356

Leica UK Ltd.
Davy Avenue
Knowlhill
Milton Keynes MK5 8LB
U.K.
Tel: 01908 666663
Fax: 01908 246246
Email: francesca.salmasi@leica-microsystems.com

Medicell International (London)
239 Liverpool Road
London N1 1LX
U.K.
Tel: 0171 607 2295
Fax: 0171 700 4156
Email: all@medicell.co.uk

Merck Ltd.
Hunter Boulevard
Magna Park
Lutterworth
Leics.
U.K.
Tel: 0800 223344
Fax: 01455 558586

Molecular Probes Europe BV
PoortGebouw
Rijnsburgerweg 10
23333 AA Leiden
The Netherlands
Tel: +31-71-5233378
Fax: +31-71-5233419
Email: eurorder@probes.nl

Morgan Matroc Ltd.
Bewdley Road
Stourport on Severn
Worcestershire DY13 8QR
U.K.
Tel: 01299 827000
Fax: 01299 827872

Nikon UK Ltd.
Instrument Division
Nikon House
380 Richmond Road
Kingston upon Thames
Surrey KT2 5PR
U.K.
Tel: 0181 541 4440
Fax: 0181 541 4584
Email: instrumentsales@nikon.co.uk

Prolume Ltd.
1085 William Pitt Way
Pittsburgh PA 15328
U.S.A.
Email: BBryanmd@aol.com

Sigma Aldrich Company Ltd.
Fancy Road
Poole
Dorset BH12 4QH
U.K.
Tel: 0800 373731
Fax: 0800 378785

Electrophysiological Measurements Using Ca^{2+}-Sensitive Microelectrodes

ROGER C. THOMAS

Introduction

Ca^{2+}-sensitive microelectrodes generate a voltage proportional to the logarithm of the calcium activity (pCa) at their tips. In principle they are simple to understand and calibrate, but difficult to make and use intracellularly. Their main advantages over other methods of measuring pCa_i (such as fluorescent dyes and luminescent proteins) are that they measure pCa at one point and do not buffer or bind the ions they measure. The apparatus required is also relatively simple and inexpensive, and the data produced are simple to process. They are also sensitive over a very wide range, from pCa 1 to at least 8.

Disadvantages of Ca^{2+}-sensitive microelectrodes are that they are difficult to construct, have very high electrical resistances, rather short lives and are likely to injure the cell membrane. They can only be used inside large cells or extracellularly. Indeed for measuring pCa just outside a cell they are unrivalled.

Principles of measurement

A glass microelectrode is made Ca^{2+}-sensitive by plugging its tip with a solution of a Ca^{2+} ligand and additives in a PVC-gelled organic solvent. Connected to a very high impedance amplifier this electrode will generate a voltage with reference to an ion-insensitive microelectrode usually filled with KCl. The voltage changes by 29 mV for a tenfold change in Ca^{2+} activity (i.e. 29 mV per unit change in pCa at 20 °C, as derived by the Nernst Equation). When made carefully these electrodes will give a lin-

Roger C. Thomas, University of Cambridge, Department of Physiology, Downing Site, Cambridge, CB2 3EG, UK (*phone* +44-1223-333869; *fax* +44-1223-333840; *e-mail* rct26@cam.ac.uk)

ear response down to a Ca^{2+} level of 10 nM or less (pCa 8 or more).

To measure intracellular Ca^{2+} the tips of both Ca^{2+}-sensitive and reference microelectrodes must be placed in the same cell. This requires a robust preparation, micromanipulators etc., special amplifiers and a Faraday cage to minimise electrical interference as shown in Fig. 1. Flowing solutions of known pCa can be used to calibrate the electrode after a successful experiment.

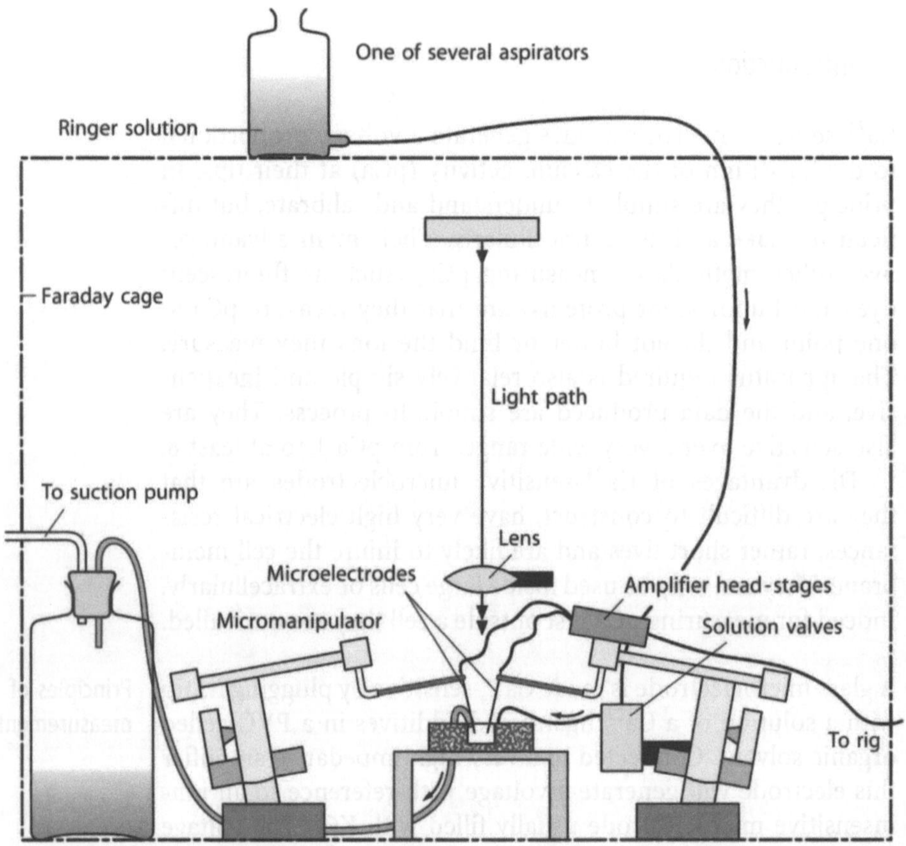

Fig. 1. Diagram showing the main components of a typical electrophysiological setup for use with Ca^{2+}-sensitive microelectrodes. The main amplifiers, stimulation, solution controller, and data display and collection system are not shown; they normally occupy an electronics rack to the right. The horizontally-mounted dissecting microscope is not shown either, nor the front of the Faraday cage.

Outline

The procedure followed in a typical experimental day is as follows:

1. Make a batch of micropipettes with tips of at least 1 μm in diameter. It may be necessary to break their tips to between 1 and 2 μm diameters.
2. Silanise the micropipettes with hot silane vapour in an evacuated heated tube.
3. Backfill several micropipettes with an aqueous reference solution and suck up a short column of the Ca^{2+} -sensitive liquid into each tip.
4. While the electrodes stabilize, set up the experimental preparation in an electophysiological rig.
5. Mount a Ca^{2+} -sensitive and reference microelectrode in the experimental chamber and once voltage outputs are stable check their responses to a change in Ca^{2+} in the superfusing solution.
6. Micromanipulate the two electrodes into a suitable cell. Apply air pressure to the back of the Ca^{2+} -sensitive electrode to maximise its response to low Ca^{2+} levels.
7. At the end of the experiment, withdraw both electrodes and calibrate the response to Ca^{2+} by running through the bath calibration solutions with levels of Na^{+}, K^{+}, Mg^{2+} and Ca^{2+} close to those found intracellularly.

Materials

Consumables

- Microelectrode glass: aluminosilicate or borosilicate glass tubing, medium-walled 1.5mm OD., obtainable from Clark Electromedical
- Ca^{2+} ligand cocktail: (all chemicals from FLUKA) (mg/ml) Ca ionophore ETH 129, 12; sodium tetrakis [3, 5-bis (trifluoromethyl) phenyl] borate, 6; 2-nitrophenyloctylether, 200; high molecular weight PVC, 33; tetrhydrofuran, 749.

This solution will remain usable for months; if it gels it can be liquified by adding more tetrahydrofuran.

- Backfill solution: 10 mM KCl and 0.01 mM $CaCl_2$ in distilled water
- Calibration solutions: may be made up using an adaptation of the following recipe for use with snail neurones: KCl 100 mM, NaCl 5 mM, $MgCl_2$ 1 mM, HEPES 10 mM, BAPTA 4 mM. Adjust pH to 7.4 by adding KOH and pCa to 6, 7 and 8 by adding $CaCl_2$. The pH may be measured with a laboratory pH electrode and the pCa with a Ca^{2+}-sensitive macro-electrode or microelectrode made as described below. The Ca^{2+}-sensitive microelectrode must first be calibrated with known Ca^{2+} buffers. McGuigan, Luthi and Buri (1991) give detailed instructions on preparing calcium buffer solutions, or they may be purchased from WPI.
- Silanisation: I recommend bis (dimethylamino)-dimethylsilane (Munoz, Deyhimi & Coles, 1993)

Equipment

- Microelectrode puller: custom made (see Thomas, 1978); but many are available commercially. A machine for quartz micropipettes is made by Sutter.
- Silanisation oven – custom made from simple components, see Fig 2.
- Micropipette filling setup. This comprises a horizontally-mounted compound microscope (objective lenses X5, X10, X20 and X40) with micromanipulators each side.
- Electrophysiological Setup: This is essentially a standard intracellular recording system with augmented shielding to protect the Ca^{2+} -sensitive potential from electrostatic interference as sketched in Fig 1.
- High input-impedance amplifiers, that for the Ca^{2+}-sensitive microelectrode having an input impedance greater than 10^{14}ohms, and an input bias current less that 10^{-14}amps. Cairn Research make a model with particularly small headstages.

Procedure

Procedure to construct Ca^{2+}-sensitive microelectrodes

1. Cut 1.5 mm O.D. aluminosilicate glass tubing into lengths 75 mm long. Fire-polish the cut ends. Pull into micropipettes with tips 1-2 µm in diameter. It may be necessary to pull sharper tips and then break them by touching a glass surface under microscopic control.
2. Place a batch of freshly-pulled micropipettes on a tray and place inside the silanisation oven as shown in Fig. 2.
3. Evacuate oven and heat it to about 400 °C in the region of the micropipette tips for 5-10 minutes.
4. Seal the oven and inject 4 µl of bis-(dimethyl)-(dimethylamino)-silane (Munoz, Deyhimi, and Coles, 1983) with a needle through the rubber seal on one end. Leave for about 5 minutes to cure.
5. Again open the tube to the vacuum pump, reevacuate oven (the vacuum inside should not have been significantly changed by the silane injection) and switch off heating current. After 5-10 minutes remove silanised micropipettes. They can be stored some days in a dessicator, although are best used within a few hours of silanization.
6. Mount a micropipette in a manipulator in the filling setup. This should hold the micropipette vertically tip down with the tip in the field of view of a compound microscope with x5, 10 and 20 objectives.

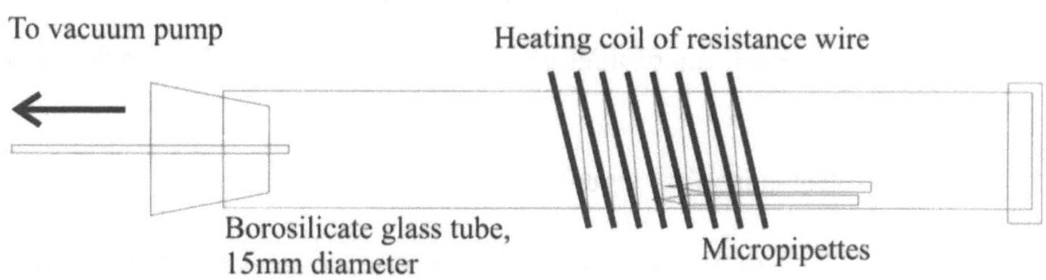

Fig. 2. Vacuum oven (Thomas, 1994) for silanizing micropipettes. The borosilicate glass tube is about 15mm diameter and 20 cm long. The left end is closed with a rubber bung, the right with a rubber cap which is the piston from a 20 ml disposable syringe.

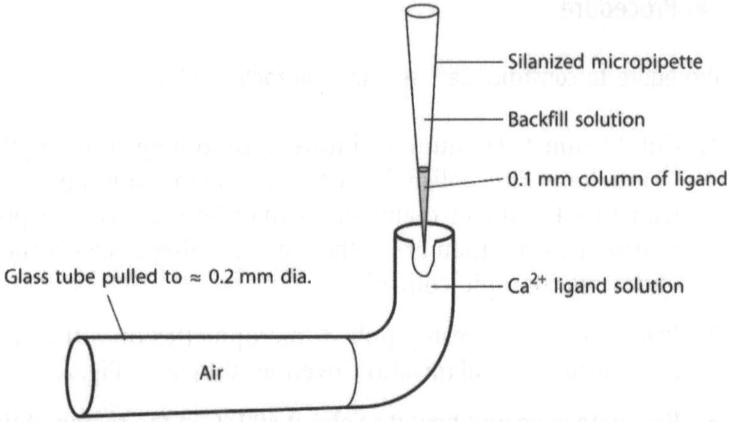

Fig. 3. Diagram of the arrangement used to fill the micropipette tips with Ca^{2+}-sensitive cocktail. The glass tube is pulled from a length of disposable micropipette tubing and bent to give a right angled tip. This allows the micropipette tip to be observed at high magnification while being filled.

7. Backfill the micropipette via a plastic tube with the backfilling solution. Connect the back of the micropipette to a 10 ml syringe full of air via a silicone rubber tube and apply manual pressure to force solution to the tip. If it runs to the tip easily silanisation may be inadequate. If the tip is blocked the glass may have been too wet before silanization.
8. Dip the micropipette tip into a small volume of cocktail held in the end of a glass tube, as shown in Fig. 3. The cocktail remains fluid for only a few minutes since the tetrahydrofuran evaporates rapidly. Allow cocktail to flow in to a column length of 100-120 µm. If necessary apply suction or pressure to adjust column length to 100-120 µm.
9. Keep micropipette tip in air for at least an hour before use.

Measuring Ca^{2+} inside a cell

1. Mount both Ca^{2+}-sensitive and reference microelectrodes over the cell chosen. Allow the voltages generated to stabilise; set the amplifier outputs to zero. (It may be wise at this point to run a low Ca^{2+} solution through the experimental

chamber to be sure the electrode gives the expected response).

2. Move the Ca^{2+}-sensitive microelectrode down to penetrate the cell. This may require a considerable movement, since the tip is unusually large. Once the electrode is inside the cell, the voltage should register a decrease of well over 150 mV. Next push the KCl-filled reference electrodes into the same cell. The potential recorded by this electrode should be subtracted from that of the Ca^{2+} -sensitive microelectrode. Applying about 1 bar pressure to the back of the Ca^{2+} -sensitive microelectrode may improve its response.

3. It may take many minutes for the electrode to record a stable $aCa_{i.}$ As shown in Fig. 4, caffeine application or depolarisation of the cell may be used to increase aCa_i, and the membrane potential may be changed to test its quality of the penetration.

4. Finally both electrodes should be withdrawn from the cell, and the Ca^{2+}-sensitive microelectrode calibrated by flowing calibration solutions through the bath.

Results

Figure 4 illustrates an experiment to measure the effects of caffeine and changes in membrane potential on pCa_i (shown as the potential generated by the Ca^{2+}-sensitive microelectrode minus that generated by the Em electrode) in a voltage-clamped snail neurone. In my experience considerable vertical movement of the Ca^{2+}-sensitive microelectrode is often required before a high and convincing pCa_i can be recorded. In this experiment the initial stable level of the Ca^{2+}-sensitive microelectrode potential was only -120 mV. Then I applied pressure and adjusted the electrode position: the recorded potential increased to over −140 mV; indicating a pCa_i of about 7, or $[Ca^{2+}]_i$ of 100 nM.

Subsequent changes in membrane potential and an application of caffeine had the expected effect on pCai, which as shown by the calibration after electrode withdrawal increased beyond pCa 7 after caffeine.

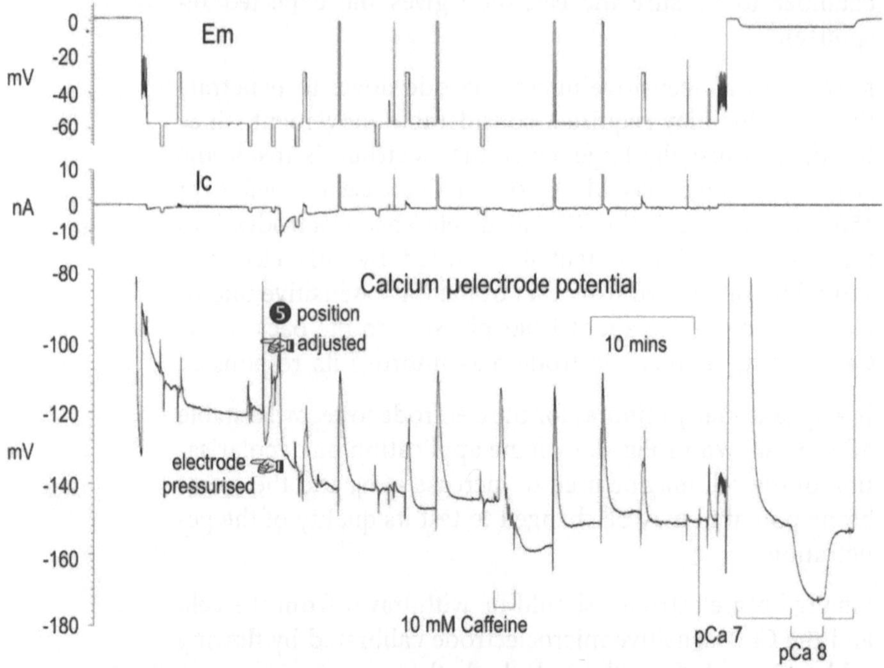

Fig. 4. Measurement of pCa_i in a snail neurone. All voltages set to zero with the electrode outside the cell. The Ca^{2+}-sensitive microelectrode was inserted first, then the membrane potential (Em) electrode, finally a third electrode to allow the Em to be clamped. About 12 minutes after penetration the Ca^{2+}-sensitive microelectrode was pressurised, and its position adjusted in a successful attempt to improve its response. After this, depolarisation produced large transient increases in Ca^{2+}. At the end of the experiment all electrodes were pulled out of the cell and calibration solutions were run through the bath (R.C. Thomas, unpublished experiment of 1993).

Although in my experience very difficult to use intracellularly, Ca^{2+}-sensitive microelectrodes work well measuring changes in surface pCa (Schwiening et al, 1993) Fig. 5 shows a recording of pCa on the surface of a voltage-clamped neurone subject to depolarizations during superfusion with low-Ca Ringer. Clear changes in surface pCa are visible, in spite of the electrical crosstalk.

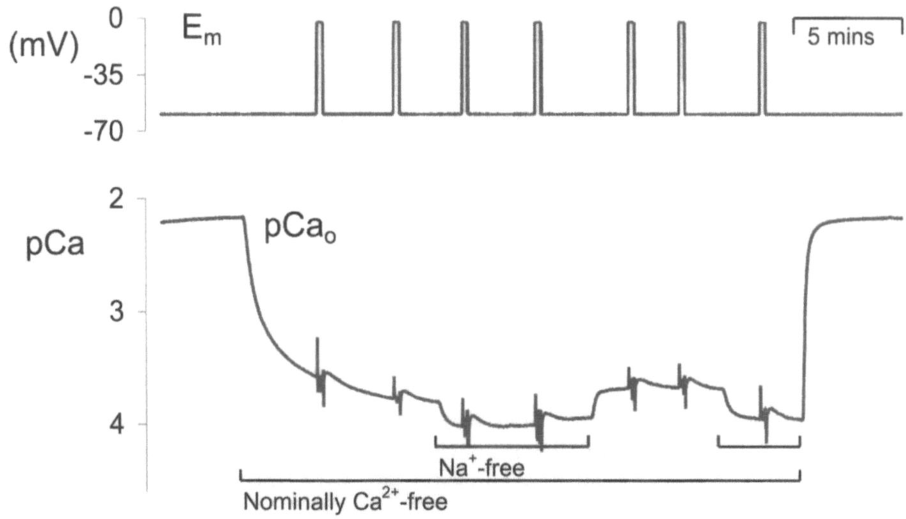

Fig. 5. Recording of pCa on the surface of a snail neucone in an isolated ganglion. The membrane potential was held at -60 mV, except where shown, by a two-electrode voltage clamp. Nominally calcium-free snail Ringer was superfused for most of the experiment. Where shown it was made Na-free. (Unpublished experiment of 1994, reproduced with permission of C.J. Schwiening)

Troubleshooting

- Poor silanization: A microelectrode may exhibit a sub-Nernstian response for several reasons, but the most likely is poor silanization. The silane used must be uncontaminated with water, and the microelectrode glass should be dry and clean. If in doubt about the quality of the silanization, check that an electrode with its tip filled with 100 µl of mineral oil or o-nitrophenyloctylether has a resistance of over 10^{12} ohms. Successful silanization seems to depend on many variables, some of which are poorly understood. Many different methods have been published, and I have tried most of them. That described above has the advantage of consuming very little silane. The apparatus apart from the vacuum pump is simple and robust.
- Inadequate amplifier insulation or input impedance. Since physiological environments are often wet, amplifier input isolation may become reduced by contamination. Test its adequacy by measuring the response to a calibration voltage

fed into the amplifier via a 10^{11}ohms resistance. Cleaning all insulation round the input may be necessary. Check that the holder for the microelectrode is also clean and dry.

- Poor sensitivity to low Ca^{2+} levels. If the response is linear down to about 10 μM but then does not change with lower levels the most likely cause is poor silanization. Failing that, it may be wise to try another batch of cocktail, or one with a different formula such as that given by Ammann (1987) or Levy & Payne (1996).

- Electrostatic interference. Extensive shielding will minimise pickup from outside the Faraday cage, but problems may arise from changing solution levels in the experimental chamber. These can be reduced by painting several mm of the shank of the microelectrode with electrically conducting paint, where it crosses the water surface.

- Sensitivity to withdrawal from cell. I have had problems with the electrode response to low levels of Ca^{2+} obviously being lost on withdrawing the microelectrode at the end of an experiment. It may be better to destroy the cell – or render it totally permeable – than pull out the electrode through an intact cell membrane. I have not myself found a way to do this, however!

- Failure to record a high pCa_i. Often the first voltage change recorded on penetrating a healthy cell is far smaller than required to indicate a normal high pCa_i. Either the electrode fails to seal in well so is exposed to Ca^{2+} entering from outside the cell (c.f. Taylor & Thomas, 1984) or its tip somehow generates a leak from internal stores. I have found it often requires a great deal of manipulation to achieve a stable recording of a believable pCa_i. Some of my best recordings have been achieved only after the tip seems to have gone right through the cell before being pulled back. Perhaps the tip needs to be inside the nucleus, away from the endoplasmic reticulum, to avoid generating leaks.

Comments

The Ca^{2+}-sensitive cocktail described above may not be the best for different applications, although I formulated it after testing many different mixtures. Other recipes are given in the Fluka handbook on Selectophores. Briefly, there are two ionophores, both invented by Professor W. Simon and his team at the Eidgenossische Technische Hochschule, or ETH, Zurich. Hence the ionophores are known as ETH 1001 and ETH 129. I have only tried the latter myself, as it is supposedly more selective. Simon Levy, another of the few who have used Ca^{2+}-sensitive microelectrodes recently, prefers ETH 1001 (Levy and Payne,1996).

Quartz tubing may well be better than either aluminosilicate or borosilicate (Munoz and Coles, 1987) but the puller (made by Sutter) is very expensive. In any case I have too little experience with quartz to be able to recommend it.

Acknowledgements. I am very grateful to the Medical Research Council for financial support, to Christof Schwiening, Helen Kennedy, Michael Rickard, and Debbie Willoughby for their help and advice and to Aileen Briggs for outstanding secretarial services.

References

Ammann D (1986) Ion-selective microelectrodes – Principles, Design and Application. Springer, Berlin, 346pp.

Ammann D, Buhrer T, Schefer U, Muller M, Simon W (1987) Intracellular neutral carrier-based Ca2+ microelectrode with subnanomolar detection limit. Pflugers Arch 409:223-228.

Kennedy HJ, Thomas RC (1995) Intracellular calcium and its sodium-independent regulation in voltage-clamped snail neurones. J. Physiol. 484:533-548.

Kennedy HJ, Thomas RC (1996) Effects of injecting calcium-buffer solutions on $[Ca^{2+}]_i$ in voltage-clamped snail neurons. Biophys. J. 70:2120-2130.

Levy S, Payne R (1996) Limulus ventral photoreceptors contain two functionally dissimilar inositol triphosphate-induced calcium release mechanisms. J.Photochem Photobiol B Biology 35:97-103

McGuigan JAS, Luthi D, Buri A (1991) Calcium buffer solutions and how to make and use them: A do it yourself guide. Can. J. Physiol. Pharmacol. 69: 1733-1749.

Munoz J-L, Deyhimi F, Coles JA (1983) Silanization of glass in the making of ion-sensitive microelectrodes. J. Neuroscience Methods 8:231-247.

Munoz J-L, Coles JA (1987) Quartz micropipettes for intracellular voltage electrodes and ion-selective microelectrodes. J.Neurosci. Methods 22: 57-64.
Schwiening CJ, Kennedy HJ, Thomas RC (1993) Calcium-hydrogen exchange by the plasma membrane Ca-ATPase of voltage-clamped snail neurons. Proc. R. Soc. Lond. B. 253: 285-289.
Taylor PS, Thomas RC (1984) The effect of leakage on microelectrode measurements of intracellular sodium activity in crab muscle fibres. J. Physiol. 352: 539-550.
Thomas RC (1978) Ion-sensitive intracellular microelectrodes: how to make and use them. Academic Press, London.
Thomas RC (1994) Current contruction protocol for calcium-sensitive intracellular microelectrodes J Physiol. 476:9P.

Suppliers

Cairn Research Ltd, Unit 3G, Brents Shipyard Industrial Estate, Faversham, Kent, ME13 7DZ, UK. Tel: +44 (0)1795 590140, fax: +44 (0)1795 590150, sales@cairnweb.com

Clark Electromedical Instruments c/o Harvard Apparatus Ltd., Fircroft Way, Kent TN8 6HE. Tel: +44 (0)1732 864001, fax: +44 (0)1732 863356, harvard@twins.co.uk

Fluka Chemie AG, CH-9470 Buchs/Switzerland Tel: 081 755 2511 Fax: 081 756 5449

Narishige Scientific Instruments 9-28 Kasuya 4-Chome Setagaya-Ku, Tokyo, Japan Fax 813 3308 2005

Sutter Instruments Co. 40 Leveroni Court Novato CA 94949 USA Phone (415) 883-0128 Fax 415 883 0572

World Precision Instruments International, Trade Center 175 Sarasota Boulevard Sarasota, Florida 34240-9258 USA Phone: (941) 371-1003 Fax: (941) 377-5428, wpi@wpiinc.com

Part B

Measuring Ca^{2+} Inside Organelles

Measuring Ca^{2+} in the Nucleoplasm of Intact Cells

MARISA BRINI, PAOLO PINTON, PAULO MAGALHÃES, LUISA FILIPPIN, TULLIO POZZAN, and ROSARIO RIZZUTO

Introduction

Aequorin is a 22 kD protein of the jellyfish *Aequorea victoria* that undergoes, upon binding of Ca^{2+} to three high affinity sites, an irreversible reaction in which a photon is emitted. In its active form, the protein includes a covalently bound prosthetic group (coelenterazine) that is released in the Ca^{2+} triggered reaction. In the 60s and 70s aequorin was, together with metallochromic indicators, the most widely employed probe for intracellular Ca^{2+} (Ridgway and Ashley, 1967; Ridgway et al., 1977; Allen and Blinks, 1978). Indeed, seminal observations carried out using aequorin as the probe revealed basic concepts of Ca^{2+} signalling, such as for example the oscillatory pattern of the agonist-dependent rises of cytoplasmic Ca^{2+} concentration in a variety of cell types. The drawback of this experimental approach, which caused its wide substitution with the trappable

Marisa Brini, University of Padova, Department of Biochemistry, Viale G. Colombo 3, Padova, 35121, Italy
Paolo Pinton, University of Padova, Department of Biomedical Sciences, Viale G. Colombo 3, Padova, 35121, Italy
Paulo Magalhães, University of Padova, CNR Center for the Study of Biomembranes, Viale G. Colombo 3, Padova, 35121, Italy
Luisa Filippin, University of Padova, Department of Biomedical Sciences, Viale G. Colombo 3, Padova, 35121, Italy
Tullio Pozzan, University of Padova, Department of Biomedical Sciences, Viale G. Colombo 3, Padova, 35121, Italy, University of Padova, CNR Center for the Study of Biomembranes, Viale G. Colombo 3, Padova, 35121, Italy
✉ Rosario Rizzuto, University of Ferrara, Department of Experimental and Diagnostic Medicine, Section of General Pathology, Via L. Borsari 46, Ferrara, 44100, Italy (*phone* +39-0532-291361; *fax* +39-0532-247278; *e-mail* rzr@unife.it)

fluorescent indicators developed by R.Y. Tsien (Tsien et al. 1982; Grynkiewicz et al., 1985), is that the protein does not permeate across cell membranes and thus needs to be microinjected. The technique was thus limited to large, sessile cells. The isolation of the aequorin cDNA (Inouye et al., 1985) completely changed this perspective. Indeed, aequorin can now be utilised in all cell types amenable to transfection, i.e. most cells. The choice between different types of indicators is now based on their different properties, as we will discuss here, and conditions can be found in which either the dyes or the photoprotein are the method of choice. Moreover, as we will discuss in this chapter and in that on the monitoring of mitochondrial Ca^{2+} concentration, specific targeting sequences can be added to the heterologous polypeptide, and thus Ca^{2+} probes with highly specific intracellular localisation can be designed.

In this chapter we first briefly review the advantages and disadvantages offered by the steep dependence of luminescence on $[Ca^{2+}]$ and the instruments needed to obtain reliable measurements, then we will present the principles, and uses, of nuclear targeted aequorin chimeras.

Aequorin as a Ca^{2+} probe

Aequorin, as produced by various *Aequorea* species, includes an apoprotein and a covalently bound prosthetic group (coelenterazine). The Ca^{2+}-dependent luminous reaction requires both the protein and the prosthetic group (Fig.1A). Since recombinant expression yields only the polypeptide, the prosthetic group must be added exogenously. This process (the "reconstitution" of the active protein) is a critical step in the use of the

→

Fig. 1. A Schematic model of the irreversible reaction of aequorin. When Ca^{2+} ions bind to the EF-hand binding sites of the reconstituted aequorin (the prosthetic group, coelenterazine, is indicated) a photon is emitted and aequorin is irreversibly discharged.
B $[Ca^{2+}]$ response curve of recombinant expressed aequorin. L, light emission immediately after exposing the recombinant photoprotein to solution with known $[Ca^{2+}]$ solution; L_{max}, integral of aequorin counts from the mixing to the end of the experiment (i.e. after aequorin consumption with excess Ca^{2+}), for details see text.

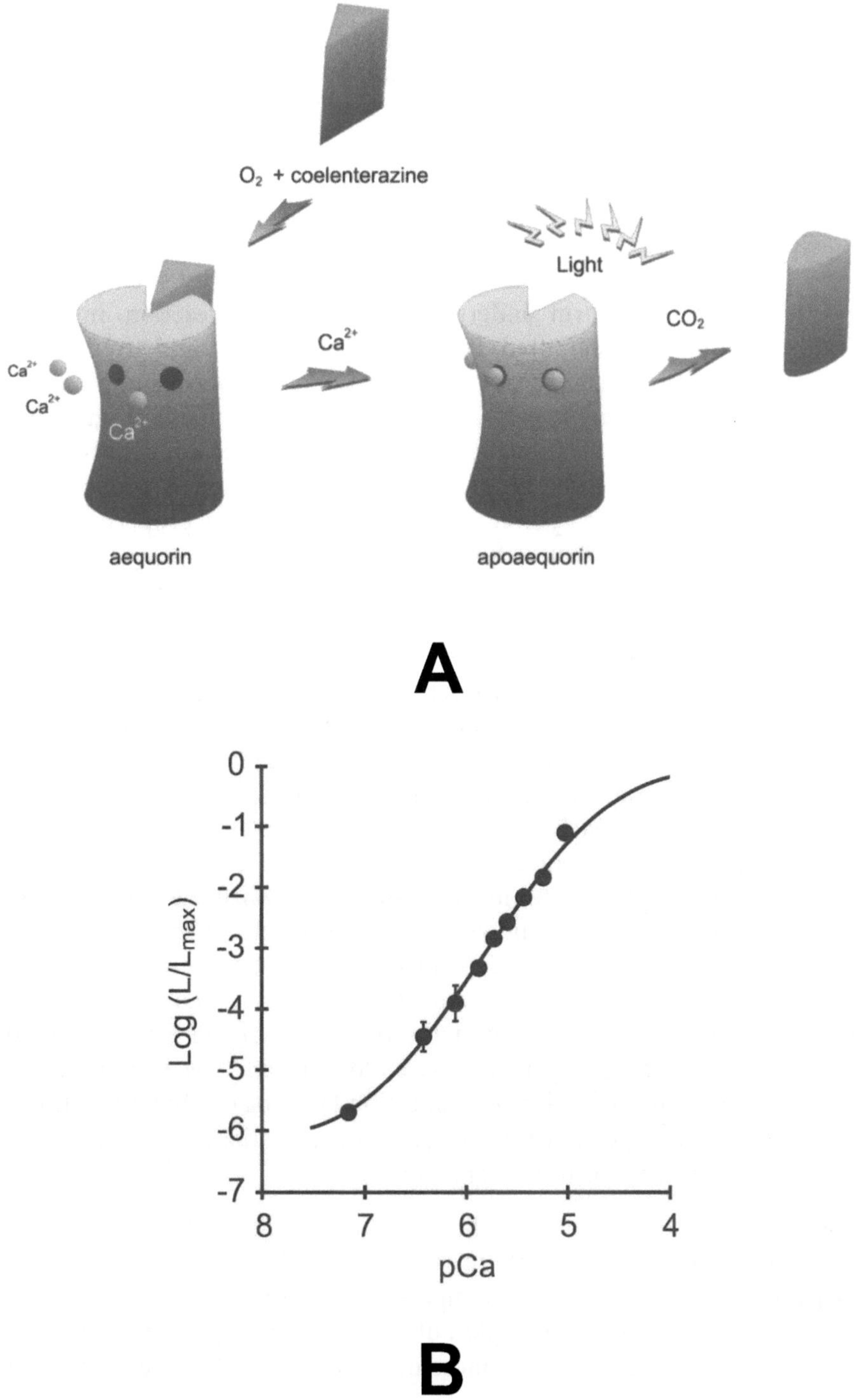

O2 + coelenterazine
Light
Ca2+
CO2
Ca2+
Ca2+
Ca2+
aequorin
apoaequorin
A
Log (L/Lmax)
0
-1
-2
-3
-4
-5
-6
-7
8
7
6
5
4
pCa
B

photoprotein. It requires diffusion of the prosthetic group across cell membrane and incorporation in the recombinant polypeptide. In our experience, both events occur quite easily in a wide variety of cell types, albeit relatively slowly. Thus, it is sufficient to add coelenterazine to the culture medium of the cells, and, provided aequorin is exposed to low Ca^{2+} and thus consumption is limited, enough active photoprotein is formed in 1-2 hours to carry out the experiment.

When Ca^{2+} ions bind to three high-affinity sites (EF-hand type), aequorin undergoes an irreversible reaction, in which a photon is emitted. For $[Ca^{2+}]$ between 10^{-7} and 10^{-5} M there is a relationship between the fractional rate of consumption (i.e. L/L_{max}, where L_{max} is the maximal rate of discharge at saturating Ca^{2+} concentrations) and $[Ca^{2+}]$. Due to the cooperativity between the three binding sites, light emission is proportional to the 2nd-3rd power of $[Ca^{2+}]$ (Fig.1B); this property on the one hand accounts for the excellent signal to noise ratio of aequorin, on the other may significantly affect the measurements (see below).

Given that the probe (differently from fluorescent indicators) is gradually consumed throughout the experiment, the signal tends to decrease and the conversion into Ca^{2+} concentration can be obtained only at the end of the experiment, when the total aequorin content is estimated and L/L_{max} can be back-calculated for each data point. The photoprotein consumption that occurs when aequorin is exposed to high Ca^{2+} concentration can become a problem in two instances: a) when the compartment or cell domain is endowed at rest with high Ca^{2+} concentrations, and extensive Ca^{2+}-dependent consumption parallels the process of reconstitution, and thus little, or negligible, functional aequorin is available at the beginning of the experiment; b) when the compartment is endowed with low resting $[Ca^{2+}]$ (and thus effective reconstitution takes place), but after stimulation large increases in $[Ca^{2+}]$ concentration, and thus quantitative aequorin discharge, occur.

The advantages of aequorin

1. Selective intracellular distribution. The main reason of the renewed interest in aequorin is that, being a protein, it can be molecularly engineered to induce its specific localisation to the cell region of interest. This approach is favoured by the recent advances in cell biology, which have revealed

the minimal targeting sequences driving endogenous proteins to their correct intracellular location. Indeed, the sequences as well as details on the molecular machinery involved have been identified for organelles, such as the mitochondria (Hartl et al., 1989) and the nucleus (Kalderon et al., 1984; Dingwall and Laskey, 1991). In other cases, much less information is available, and thus it is necessary to fuse the heterologous protein to the entire resident polypeptide.

2. Wide dynamic range. As to the sensitivity range, aequorin is well suited for measuring [Ca^{2+}] between 0.5 and 10 μM. The wide dynamic range, together with the low buffering effect (see below), makes the photoprotein the tool of choice when it is desirable to estimate quantitatively the large [Ca^{2+}] rises that occur in some cell types (e.g. those occurring in neuronal cells). Also this wide dynamic range, however, is insufficient when, thanks to specifically targeted chimeras, the photoprotein is addressed to compartment endowed with high [Ca^{2+}], such as the organelle acting as intracellular store of agonist-releasable Ca^{2+}. In this case, strategies can be devised to reduce the affinity of the photoprotein. The first is to act on the Ca^{2+}-binding sites of the photoprotein. The Asp119→Ala mutation reduces the affinity of the photoprotein approx. 50 fold (Kendall et al., 1992), and has been instrumental for the development of the aequorin probes targeted to the endoplasmic reticulum (ER) (Montero et al., 1995), the sarcoplasmic reticulum (SR) (Brini et al., 1997), the Golgi apparatus (Pinton et al., 1998) and the subplasmalemmal space (Marsault et al., 1997). No other mutation further reducing the affinity of the photoprotein has been reported. The second point of action is the prosthetic group. Numerous chemical modifications of the prosthetic group have been carried out, which modify, in different ways, the Ca^{2+}-triggered reaction of the photoprotein (Shimomura et al., 1989). Among these, particularly useful have been those (coelenterazine i and n) that reduce light emission at high [Ca^{2+}] and thus, in combination with the apoprotein mutation, have made feasible the direct measurement also in the organelles endowed with Ca^{2+} concentrations approaching the millimolar range (Barrero et al., 1997). Finally, the last solution to the experimental problem is substituting

Ca^{2+} in the cells with a cation with similar physiological behaviour, but less effective in eliciting the luminous reaction of aequorin. Such a cation is, fortunately, a classical surrogate of Ca^{2+} in physiology studies, Sr^{2+}, that permeates across Ca^{2+} channels and is transported by the Ca^{2+} ATPases. Indeed, the approx. 100 fold lower sensitivity of aequorin for Sr^{2+} accounts for the drastically lower rate of photoprotein consumption also in the intracellular stores, and thus prolonged and reliable monitoring of these key cell domains can be achieved (Montero et al., 1995).

3. High signal-to-noise ratio. Mammalian cells are not endowed with chemiluminescent proteins, so the background of aequorin measurement is very low (typically 4-6 cps from a 13 mm coverslip of transfected cells). Moreover, the steep relationship between the increases in light emission and Ca^{2+} concentration (see above) accounts for the very large luminescence peaks observed upon stimulation of cells. For an average level of transfection, the total number of detectable photons from 10^3-10^4 cells is approx. 10^7-10^8, and peaks increases of $>10^3$ and 10^4 cps, i.e. >100-1000 fold over background, can be detected with cytosolic and mitochondrial aequorin, respectively.
4. Low Ca^{2+} buffering effect. Thanks to the excellent signal to noise ratio, reliable aequorin measurements can be obtained with moderate levels of expression. Indeed, although all probes perturb Ca^{2+} homeostasis because they bind Ca^{2+} and thus act as Ca^{2+} buffers, this effect is much less relevant for aequorin than for the trappable fluorescent dyes (typically loaded at concentrations of >50 μM, i.e. >100 times higher than recombinant aequorin) (Brini et al., 1995).
5. Possibility of co-expressing the Ca^{2+} probe with a protein of interest. The possibility of modifying the molecular repertoire of a cell is today one of the most powerful tools for dissecting complex signalling pathways, unravelling the relative contribution of the different components. This is certainly true also for calcium signalling, and one of the experimental tasks we often face is that of measuring Ca^{2+} concentration specifically in cells expressing a normal or mutated signalling component. With dyes, given that it is not possible

to load the indicator in the subset of transfected cells, this is usually accomplished by either using clones stably expressing the transgene (with the problems deriving from the natural variability of cell clones) or using laborious single cell analysis of transfected cell populations. Aequorin offers an easy solution to the problem, because it can be co-expressed with the protein of interest. In transient expression studies, the Ca^{2+} probe is exclusively localised to the fraction of transfected cells (depending on the cell type 3-70% of the total population), which are thus very representative of the behaviour of the parental population (Bastianutto et al., 1995; Lievremont et al., 1997).

The disadvantages of aequorin

1. Low light emission. The major disadvantage in the use of aequorin is the low amount of light emitted by the photoprotein. Each aequorin molecule emits only one photon, and only a very small fraction of the photoprotein pool ($<10^{-3}$) emits light throughout the experiment. This is not a major problem when total recordings from a coverslip of transfected cells are carried out, as described above. However, it is often desirable to carry out single cell imaging experiments, similar to the imaging of cytoplasmic Ca^{2+} concentration with dyes (that revealed phenomena of utmost interest, such as waves, localised rises, etc.). With aequorin, this is quite difficult. Special imaging systems are needed, with enhanced sensitivity at the cost of lower spatial resolution. Thus, although single cell imaging experiments with recombinant aequorin have been carried out and provided interesting biological information (Rutter et al., 1996), this approach is technically difficult and possibly disappointing if an image quality similar to those of dyes is expected.
2. Overestimation of the average rise in cells (or compartments) with dishomogeneous behaviour. Due to the steep Ca^{2+} response curve of aequorin, if the probe is distributed between a high and a low Ca^{2+} domain, the former will undergo a much larger discharge. The total signal will be calibrated as an "average" $[Ca^{2+}]$ increase, that will be severely biased by the region at high Ca^{2+}. In other words, given that an increase in 1 pCa unit causes a 100-1000 fold increase in light emission, a tiny volume undergoing a very large $[Ca^{2+}]$

increase will drastically increase total light output and thus be "interpreted" by the calibration algorithms as a moderate Ca^{2+} rise throughout the cells. Conditions can be envisaged in which this effect can become very significant: for example, a Ca^{2+} rise of 3 μM detected in neurons could be, in fact, lower, but severely biased by hotspots occurring throughout the dendritic tree.

3. The loading procedure. The obvious requirement of this approach is that a cell is amenable to transfection. The significant improvement of transfection techniques has made this problem less severe. In our experience, all cell types can be transfected, with either the simple calcium phosphate procedure (which proves effective in most cell types) or other techniques (liposomes, gene gun, electroporation). Conversely, the need to maintain the cells in culture for enough time to produce the recombinant protein can be an important limitation, rendering the extension of the approach to interesting cell types (e.g. pancreatic acinar cells, spermatozoa, etc.) quite difficult.

The targeting strategy of selectively localised aequorins

Selective targeting of recombinant polypeptides is a major issue of modern cell biology. The machinery and the signals that are responsible for the selective localisation of endogenous proteins have been well characterised in some cases (e.g. mitochondria) and in others are largely unknown. In the case of aequorin we have adopted three different strategies. Before describing them it is worth mentioning that chimeric aequorins with long extension at the N-terminus can be constructed without appreciable alterations in the chemiluminescent properties and Ca^{2+} affinity of the probe. On the contrary even small modifications at the C-terminus result in either irreversible loss of chemiluminescence or in dramatic increases in the rate of Ca^{2+} independent luminescence which poses big problems in quantification of the signal (Nomura et al., 1991).

As mentioned above three strategies have been conducted:

1. Addition at the N-terminus of aequorin of a known targeting sequence. This has been the case of aequorin targeted to mitochondria (Rizzuto et al., 1992), to the nucleus (Brini et al., 1993; 1994) and to the Golgi (Pinton et al., 1998).
2. Fusion of aequorin at the C-terminus of a protein that contains its own targeting strategy. In this latter case it is important that the C-terminal extension (due to the aequorin polypeptide) does not affect the targeting by itself. By this approach we have constructed the aequorins targeted to the sarcoplasmic reticulum (SR) (Brini et al., 1997), to the plasma membrane (Marsault et al., 1997) and to the intermembrane space of the mitochondria (Rizzuto et al., 1998).
3. Fusion of aequorin at the C-terminus of a polypeptide that binds firmly to an endogenous protein. This is the strategy adopted for the targeting to the ER lumen (Montero et al., 1995). It should be stressed that efficient and selective targeting is an essential part of the whole measurement, given that the spatial resolution of the methodology depends not on sophisticated hardware of subcellular imaging, but rather on the selective targeting of the probe. Thus even a small fraction of missorted probe can profoundly affect qualitatively and quantitatively the measurement. It is thus important that the subcellular localisation is checked with most accuracy. For this purpose both immunohistochemical approaches (at the optical and EM level) as well as functional (subcellular fractionation, specific drugs) have been employed.

Nuclear targeting

Principles

The intracellular membrane systems of eukaryotic cells act as a barrier that divides the cell into functional compartments. Even when these compartments are established and maintained there are fluxes and exchanges of information between them.

In particular, there is a continuous exchange of macromolecules between the nucleoplasm and the cytoplasm. The cell nucleus is surrounded by the nuclear envelope which is composed of the outer and the inner nuclear membranes separated by the perinuclear space. The nuclear membrane is punctured by 2-60 pores/μm^2. Transport of macromolecules in both directions occurs through nuclear pore complexes. They provide aqueous channels with a functional diameter of 9-10 nm, allowing diffusion of ions, small molecules, and proteins with molecular masses up to 40-50 kDa (Peters, 1986). The transport of large proteins is an active process and it is temperature dependent, it requires ATP and displays a saturation kinetic.

Proteins are targeted to the nucleus by specific signals within the proteins' primary sequence. Our current view of the nuclear import process derives from the discovery of nuclear localisation sequences, NLSs, first suggested by De Robertis and colleagues (De Robertis et al., 1978). A sequence can be defined by nuclear localisation signal when it can by genetic or biochemical fusion render a cytoplasmic protein nuclear or when deleted or mutated, no longer promote nuclear uptake of the protein in which it resides. Numerous similarly defined sequences have been identified in nuclear proteins. Comparisons of these sequences have led to the proposal that the SV40 T antigen NLS is the prototypic NLS. In general they share common characteristics:

- typically short sequences usually not more than 8-10 amino acid
- contain a high proportion of positively charged amino acid (lysine and arginine)
- are not located in a specific site of a protein
- are not removed following the localisation
- can occur more than once in a given protein

Some proteins do not possess their own NLS and enter the nucleus via cotransport with another protein (Dingwall et al., 1982; Zhao and Padmanabhan, 1988). In each case at least one NLS is necessary for import of the protein complex. The possibility exists that proteins smaller than estimated diameter of the aqueous channel may simply enter the nucleus by diffusion. Experiments with variously sized dextrans and proteins microinjected into the cytoplasm have shown that molecules smaller than 60 kDa can enter the nucleus by diffusion (reviewed in Peters,

1986). However certain small proteins such as histone H2B (15 kDa) contain a NLS (Moreland et al., 1987), and histone H1(21kDa) enters the nucleus in a temperature- and energy-dependent manner, suggesting active import of these small proteins (Breeuwer and Goldfarb, 1990).

In this context a simple and efficient manner to target specifically a protein to the nucleus seem to be the introduction in its sequence of a nuclear localisation signal.

The chimeras

Targeting a Ca^{2+} indicator to the nucleus appears a major advance in the selective measurement of nuclear Ca^{2+} concentration. An ideal indicator for the measurement of nuclear $[Ca^{2+}]$ should in theory respond to the following criteria: (i) be exclusively localised in the nucleoplasm; (ii) not be affected by nuclear factors, (iii) be loaded easily in intact cells; and (iv) be amenable not only to single cell analysis but also to population measurements. In the following pages we describe the construction of aequorin chimeras which can be localised either constitutively or after hormonal treatment in the nucleoplasm, and their use for monitoring $[Ca^{2+}]_n$.

Our strategy for addressing aequorin to the nucleus was based on the targeting sequence of the rat glucocorticoid receptor, GR. The GR is cytosolic, because the nuclear localisation signal (NLS) present in its sequence is normally masked, presumably by the interaction with a heat-shock protein. Upon hormone binding, the inhibition is relieved and translocation into the nucleus occurs (Picard and Yamamoto, 1987). When a foreign protein is fused to the GR, it is transported to the nucleus when the cells are treated with glucocorticoids. If however, a smaller portion of the GR, lacking the hormone-binding domain, is used for the chimera, the NLS is relieved from hormone-dependence and the polypeptide is costitutively localised in the nucleus (Rusconi and Yamamoto, 1987).

We have constructed two aequorin chimeras by fusing the photoprotein to two different portions of the GR (Fig. 2).

In order to obtain constitutive targeting to the nucleus, a smaller part of the rat glucocorticoid receptor cDNA, including the DNA binding domain and the nuclear localisation signal NL1 (Picard and Yamamoto, 1987) was fused in frame with the aequorin cDNA (nuAEQ). In order to prevent interference with the luminescence properties of aequorin (Nomura et al.,

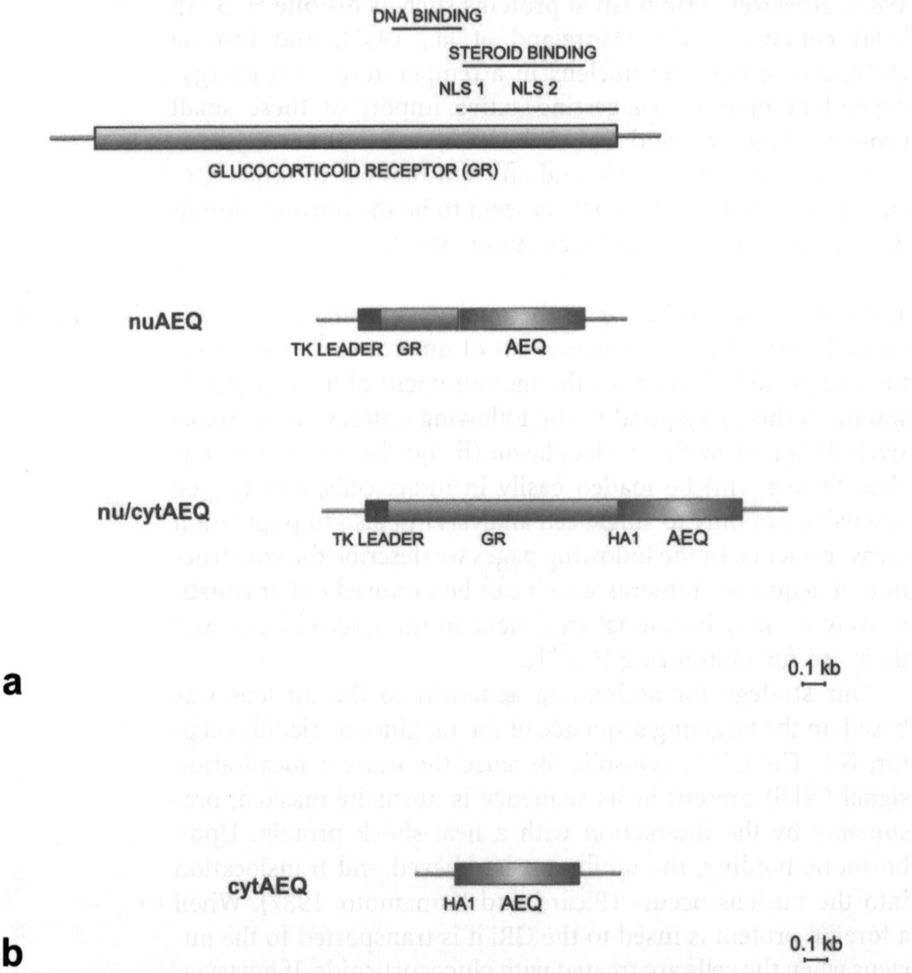

Fig. 2. Schematic maps of the "nuclear" and cytosolic aequorin chimeras. On the top, the wild type glucocorticoid receptor (GR) is shown: the positions of the nuclear localisation signals (NLSs) and of the DNA and steroid hormone binding domains are marked with lines over the cDNA. Below, the nuAEQ, the nu/cytAEQ (**a**) and the cytAEQ (**b**) chimeras are shown. The GR moiety is aligned with its position in the wt GR cDNA.
nuAEQ: the chimeric cDNA encodes a fusion protein composed by a portion of the glucocorticoid receptor GR (aa 407-524), including the nuclear localisation signal NLS1 and the untagged aequorin. This polypeptide is constitutively localised in the nucleus. nu/cytAEQ: the chimeric cDNA encodes a fusion protein composed by a portion of the glucocorticoid receptor GR (aa 407-794), including the nuclear localisation signals (NLSs), the hormone binding domain and the HA1 tagged aequorin. This polypeptide is localised in the cytosol in the absence of glucocorticoids, and translocated to the nucleus upon hormone treatment.
cytAEQ: no targeting sequence is added to aequorin, the sequence of aequorin was modified only by adding the epitope tag HA1.
Coding and non-coding regions of the cDNA are indicated as boxes and lines, respectively.

1991), the GR sequence was added to the N-terminus of the protein. Essentially the same portion of GR (aa 407-545) has previously been shown to mediate constitutive nuclear localisation of β-galactosidase (Picard and Yamamoto, 1987).

We then constructed a second chimera (nu/cytAEQ), which differs from nuAEQ in two respects: (i) a 9 amino acid sequence, corresponding to the HA1 epitope of hemagglutinin (Field et al., 1988), was added to the N-terminus of the aequorin molecule, thus providing a strong immunological tag for the localisation of the protein; and (ii) the GR moiety of the chimera is longer (amino acids 407-794 of wild type GR), thereby including, together with the NLSs (NL1 +NL2), also the whole hormone binding domain (HBD). An obvious prerequisite for the use of this chimera as a selective nuclear Ca^{2+} indicator is that the recombinant protein shuttles between nucleus and cytosol upon treatment with glucocorticoid hormones, and is indeed entirely localised in the nucleus in the appropriate experimental conditions (see "The experimental procedure" section).

Figure 2 shows a schematic map of the nuAEQ and nu/cytAEQ cDNA constructs and the details of the construction strategy are described in the legend.

The availability of two chimeras, one permanently localised in the nucleus, the other localised in the nucleus or in the cytosol, depending on hormone addition, offers an important flexibility to the technique. The constitutive nuclear localised chimera is the probe of choice in cells which are significantly modified in their behaviour by treatment with steroid hormones; the cytosol/nuclear shuttling chimera appears most useful for determining, in the same cells and with the same probe, minor differences in [Ca^{2+}] between the two compartments.

Procedure

Cell cultures

HeLa cells were grown in Dulbecco's modified Eagle's medium (DMEM), supplemented with 10% foetal calf serum, 2 mM glutamine, penicillin (200 U/ml) and streptomicin (0.2 mg/ml) in 75 cm² flasks. For the transient expression experiments, the cells were seeded onto 13 mm cover slips and allowed to grow to 50% confluence. At this stage, transfection with recombinant plasmids (4 μg DNA/coverslip) was carried out according to the calcium phosphate procedure (Rizzuto et al., 1995).

Primary cultures of skeletal muscle were prepared from new born rats (2-3 days) as described previously (Cantini et al., 1994). The posterior limb muscles were removed and dissociated by four successive treatments with 0.125% trypsin in phosphate buffer saline (PBS). The first harvest, which contains mostly fibroblasts and endothelial cells, was discarded. The remaining cell suspension was filtered through a nylon gauze. Cells were collected by centrifugation for 10 min at 1200 rpm (Haereus Minifuge GL), resuspended in DMEM, supplemented with 10% foetal calf serum and 4.5 g/l glucose, and then plated in 10 cm Petri dishes at a density of 10^6 cells/dish to decrease the number of fibroblasts in the culture. After 1 hour incubation at 37 °C the non adherent cells were collected and seeded at a density of 3×10^5 cells onto 13 mm cover slips coated with 2% gelatin. Transfection was carried out on the second day of culture with a conventional calcium phosphate procedure. 12 h after transfection the medium was changed to DMEM supplemented with 10% horse serum to increase myoblast fusion. After 24 h, the cells were switched to a maintenance medium (2% horse serum) and kept under these conditions until used.

Primary cultures of cortical neurons were obtained from 17th days embryos. Cortexes are explanted in PBS additioned of 6 mM glucose and 1% BSA. The tissue is dissociated by treatment with papain followed by mechanical disruption. Cells are then spinned and resuspended in defined medium (2/3 DMEM, 1/3 Ham's F12 supplemented with alanine, proline, glutamate, glutamine, vitamine B12, lipoic acid, $ZnSO_4$ and 10% FCS), counted and plated on poly-L-lysine coated glass coverslip at a density of 10^6 cells/coverslip. The medium is replaced after 3 h. Transfection with calcium phosphate procedure was performed on the fourth day of culture, using 4 μg of DNA per coverslip.

Transfection We currently employ in the lab mainly 3 transfection procedures: calcium phosphate co-precipitation, electroporation and particle gun delivery. Other groups have used other standard procedures for loading aequorin cDNA into cells, including viral infection. Both transient transfection and stable clones can be generated. Overall, recombinant expression of aequorin has proven so far totally innocuous for the cells, independently of the subcellular localisation of the protein.

Culture conditions of the nu/cytAEQ chimera transfected cells

The cells are grown in the culture conditions described above until the transfection was carried out. After transfection the medium is replaced with fresh DMEM supplemented with 10% FCS. When the nu/cytAEQ is employed as nuclear Ca^{2+} probe, the cells are treated for 2-24 h (depending on the cell types) before the experiment with 10 μM dexamethasone in order to induce the translocation of the recombinantly expressed protein to the nucleus. To use the nu/cytAEQ chimera as a cytosolic Ca^{2+} probe, it is necessary to control the cultures conditions. In fact, it was previously shown that the nuclear localisation of the intact glucocorticoid receptor is partially induced by certain components in standard growth medium. In particular, the glucocorticoid receptor localisation is strongly affected by foetal calf serum and the phenol red pH indicator included in the growth medium (Picard and Yamamoto, 1987). In order to prevent the nuclear localisation of the chimera the cells were transfected in growth medium lacking phenol red and supplemented with 10% FCS treated with charcol in order to reduce the level of serum factors that induce nuclear localisation. Dexamethasone induces complete nuclear localisation even in absence of both serum and phenol red.

Reconstitution

After expression, the recombinant apoprotein must be reconstituted into functional aequorin. This can be accomplished by incubating transfected cells with the chemically synthesised prosthetic group, coelenterazine (now commercially available from Molecular Probes). In our experience, coelenterazine is freely permeable across cell membranes and reconstitution may occur within all intracellular compartments to which the photoprotein has been targeted. The reconstitution process is relatively slow (optimal reconstitution is usually observed after 1-2 hours of incubation with coelenterazine). In compartments with low Ca^{2+} (cytosol and nucleus, described in this chapter and mitochondria and mitochondrial intermembrane space, described in the chapter by Pinton and colleagues) the functional aequorin pool gradually increases during the reconstitution, as consumption is negligible. Efficient reconstitution is thus achieved by incubating, before the experiment, the cells with 5 μM coelenterazine for 2-4 h at 37 °C in DMEM supplemented with 1% FCS in 5% CO_2 atmosphere.Then the coverslip is directly transferred to the luminometer chamber, where it is perfused with KRB saline solution (Krebs-Ringer modified buffer: 125 mM NaCl, 5 mM KCl, 1 mM Na_3PO_4, 1 mM $MgSO_4$, 5.5 mM glucose, 20 mM HEPES, pH 7.4, 37 °C).

Coelenterazine is normally prepared as a stock solution in pure methanol, at concentration of 0.5 mM. 50 µl aliquots of such stock solutions are kept at –80 °C in the dark.

Measurements We have employed for most of our experiments a purpose-built luminometer (Fig. 3). In this system, the perfusion chamber, on top of a hollow cylinder, thermostatted by water jacket, is continuously perfused with buffer via a peristaltic pump. The cell coverslip is placed at a few mm from the surface of a low noise phototube. The photomultiplier is kept in a dark refrigerated box. An amplifier discriminator is built in the photomultiplier housing; the pulses generated by the discriminator are captured by a Thorn EMI photon counting board, installed in a IBM-

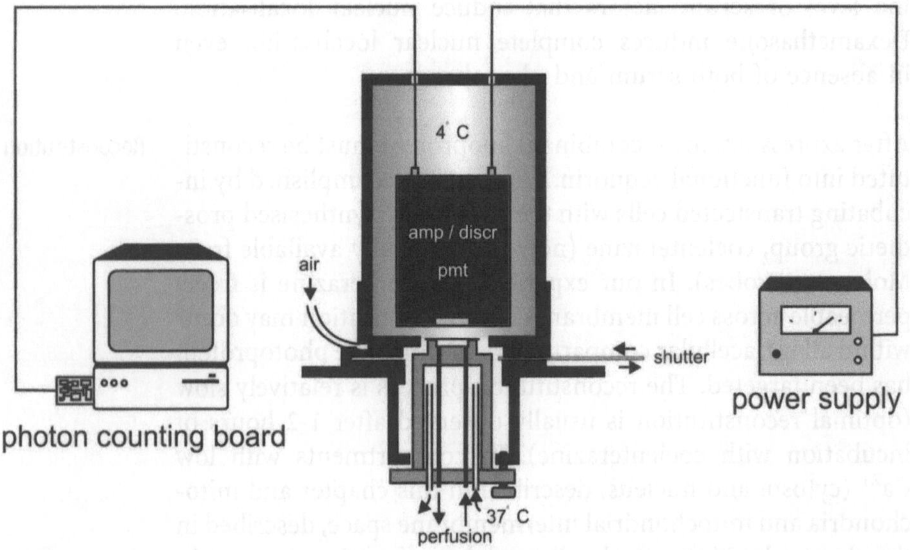

Fig. 3. The aequorin measuring system. pmt=photomultiplier; amp/discr=amplifier/discriminator. The photomultiplier is kept in a cooled dark box, the cell chamber at room temperature. During manipulations on the cell chamber, the photomultiplier is protected from light by a shutter. During aequorin measurements, the shutter is open and the chamber held in position (in close proximity of the photomultiplier window) by a O-ring. An amplifier-discriminator is built in the photomultiplier housing. The photon counting board allows the storing of the data in the computer memory for further analyses.

compatible computer. The board allows the storing of the data in the computer memory for further analyses.

In order to calibrate the raw luminescent signal in terms of $[Ca^{2+}]$ an algorithm has been developed (based on that of Blinks et al., 1978) that takes into account the instant rate of photon emission and the total amount of photon that can be emitted by the aequorin of the sample. In order to obtain the latter parameter, at the end of each experiment the cells are lysed by perfusing them with a hypoosmotic medium containing 10 mM $CaCl_2$ and a detergent (100 μM digitonin) in order to discharge all the aequorin that was not consumed during the experiment.

Results

Biological results

Although the best known targets of Ca^{2+} are located in the cytosol or in the inner surface of the plasma membrane (calmodulin, protein kinase C, ion channels, etc.) evidence has accumulated over the past few years indicating that several nuclear events (breakdown of the nuclear envelope, transcription activation, etc.) may also be modulated to some extent by changes of $[Ca^{2+}]$ in the nucleoplasm (Collart et al., 1991; Gaido and Cidlowski, 1991; Tombes et al., 1992). The issue of measurement and control of nuclear calcium has traditionally been a highly controversial topic. The nuclear envelope has been viewed by some as a structure that offers no barrier to the movements of calcium, which would thus freely and immediately equilibrate between the cytosolic and the nuclear compartments, and by others as an effective sieve that instead limits the traffic of calcium between the two compartments.

In the experiments presented in this article, three different embryological cell types (HeLa cells, rat primary cultures of muscle cells and of cortical neurons) were transiently transfected with the two described GR-aequorin chimera and with cytosolic aequorin.

Transfection, hormone treatment of the cells, as well as aequorin reconstitution and measurements, were carried out as described in the experimental procedure section.

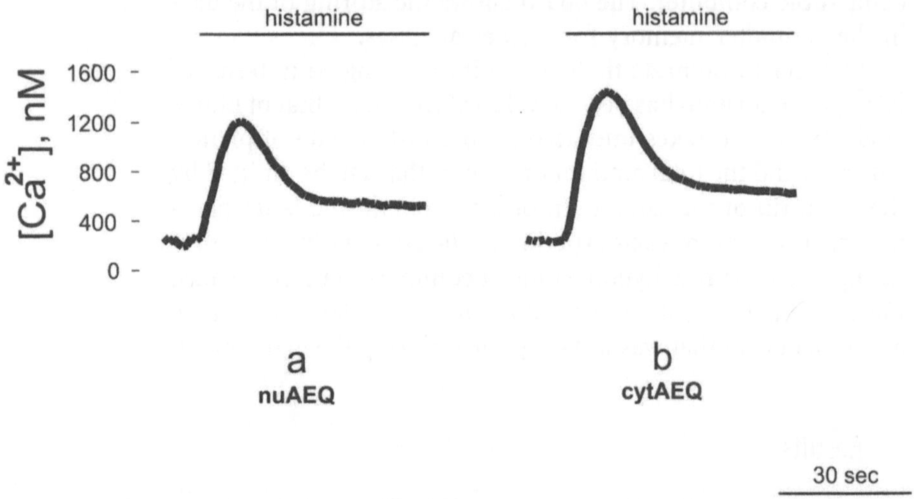

Fig. 4. Monitoring of $[Ca^{2+}]_n$ and $[Ca^{2+}]_c$ in HeLa cells with the nuAEQ and the cytAEQ chimeras. Calculated $[Ca^{2+}]_n$ (a) and $[Ca^{2+}]_c$ (b) values, from a monolayer of HeLa cells transiently expressing nuclear aequorin *(nuAEQ)* and cytosolic aequorin *(cytAEQ)*, respectively. Where indicated the cells were stimulated with 100 μM histamine, added to the KRB medium (see text).

Fig. 4a shows the monitoring of $[Ca^{2+}]_n$ in HeLa cells using the nuAEQ chimera. In this case, no epitope tag was included in the sequence and attempts to determine the subcellular localisation of the recombinant protein by immunofluorescence utilising antibodies raised against aequorin were unsuccessful. In fact no specific signal was observed either in the nucleus or in the cytosol, probably due to the low immunogenicity of aequorin. A different approach was thus used. We measured the release of nuAEQ or cytAEQ in the supernatant of HeLa cells permeabilised with digitonin, a detergent preferentially affecting cholesterol-containing membranes. The kinetics of digitonin release were strongly suggestive of nuclear localisation (Brini et al., 1993).

The luminescence data were converted into $[Ca^{2+}]$ values, based on the response curve of aequorin to Ca^{2+}, assuming that pH, Mg^{2+}, and ionic strength in the nucleus are not significantly different from those of the cytosol. A resting $[Ca^{2+}]_n$ of approximately 100-200 nM can be estimated (although it should be remembered that, because of its steep Ca^{2+} response curve, aequorin is inaccurate in measuring $[Ca^{2+}]$ below 0.3-0.4 μM, i.e.

when the signal is in the range of the background signal). Upon stimulation with the InsP3-generating agonist histamine, $[Ca^{2+}]_n$ surges to a peak approximately 1.1 μM, and then rapidly reaches a plateau of approximately 0.4 μM, which declines slowly under continuous agonist stimulation. The monitoring of $[Ca^{2+}]_c$ with cytosolic recombinant aequorin (cytAEQ) is shown, for comparison in Fig. 4b. The kinetic behaviour and the absolute $[Ca^{2+}]$ values before or after stimulation are very similar in the two compartments.

Fig. 5 shows the localisation of the expressed nu/cytAEQ chimera transfected in HeLa cells. In untreated cells (a) the photoprotein appears exclusively localised in the cytoplasm; and upon hormone treatment (10 μM dexamethasone for 2 h) the chimera nu/cytAEQ translocated to the nucleus. Immunocytochemistry analysis was carried out utilising the antibody against

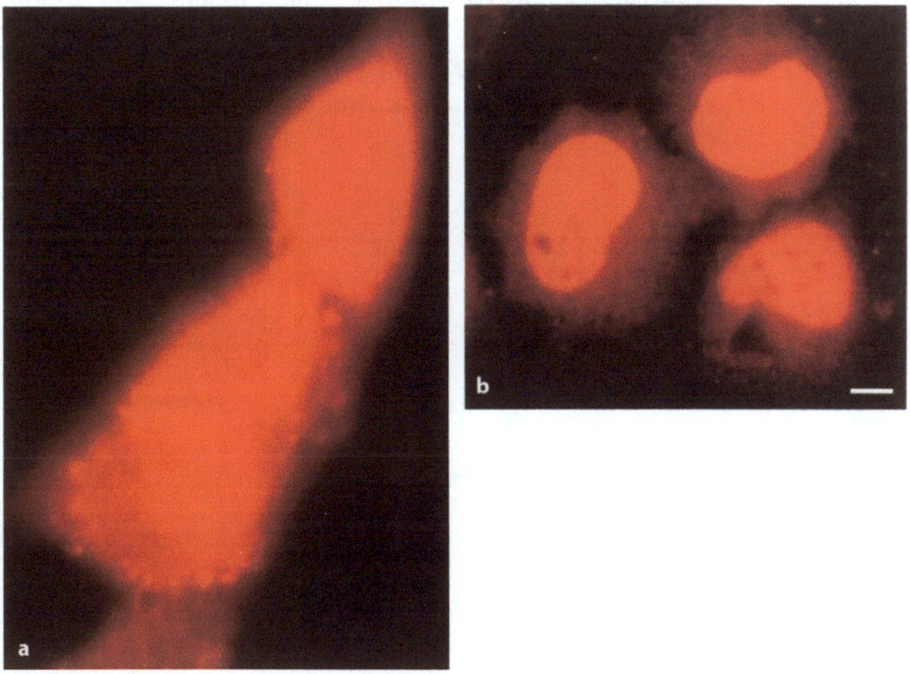

Fig. 5. Immunocytochemical localisation of the chimeric aequorins. Fluorescence microscope images of HeLa cells transfected with the nu/cytAEQ, labelled with anti HA1 mouse monoclonal antibody and finally revealed with TRITC-conjugated rabbit anti-mouse secondary antibodies. **(a)** nu/cytAEQ, **(b)** nu/cytAEQ after 2 h incubation in the presence of 10 μM dexametazone. Bar, 10 μm.

the HA1 epitope tag, as described in the chapter by Pinton and colleagues.

Fig. 6 shows the monitoring of $[Ca^{2+}]_n$ and $[Ca^{2+}]_c$ in parallel batches of HeLa cells transfected with the nu/cytAEQ chimera either treated or not treated with the hormone. $[Ca^{2+}]_n$ was approximately 200 nM at rest. After agonist stimulation, it peaked at ~1.2 μM, the same value observed in cells in which nu/cytAEQ was used as a cytosolic probe (i.e. in the absence of the hormone). These results, in good agreement with those in which the $[Ca^{2+}]$ in the nucleus and in the cytoplasm was measured with nuAEQ and cytAEQ suggest that, in terms of Ca^{2+} signalling the nucleus passively follows the cytosolic behaviour. In these experiments, the presence of the hormone does not cause per se any appreciable change in Ca^{2+} response of the cells. In fact we monitored both the $[Ca^{2+}]_c$ and the $[Ca^{2+}]_n$ with cytAEQ and nuAEQ in hormone treated and in control cells and we measured, in all the cases, similar $[Ca^{2+}]$ changes (not shown).

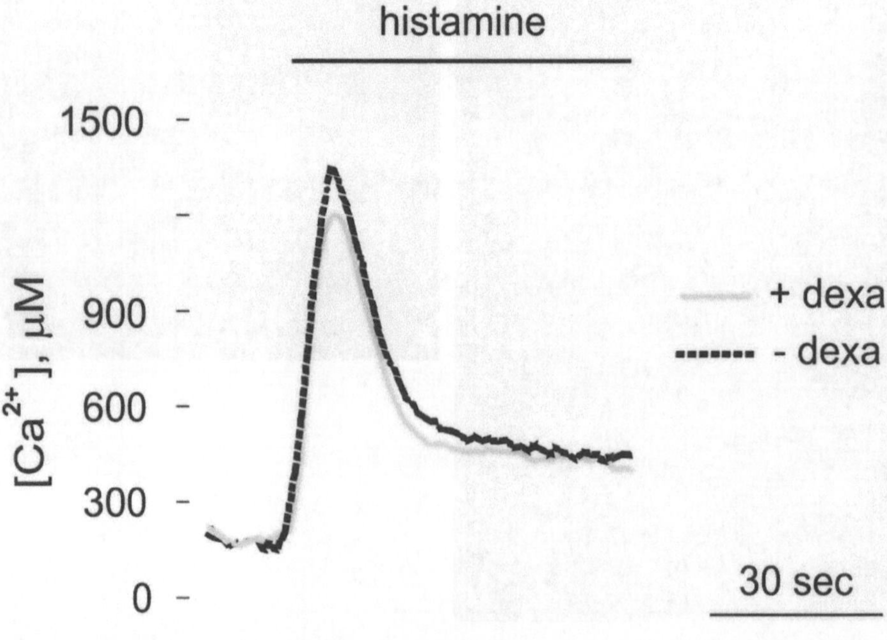

Fig. 6. Monitoring of $[Ca^{2+}]_c$ and $[Ca^{2+}]_n$ in HeLa cells with the nu/cytAEQ chimera. The figure shows the use of the nu/cytAEQ probe as nuclear probe (+dexa) or as cytosolic probe (-dexa). Where indicated the cells were stimulated with 100 μM histamine, added to the KRB medium. The dotted and solid lines refer, respectively, to control cells and to the cells incubated 2h before the experiment with 10 μM desamethasone.

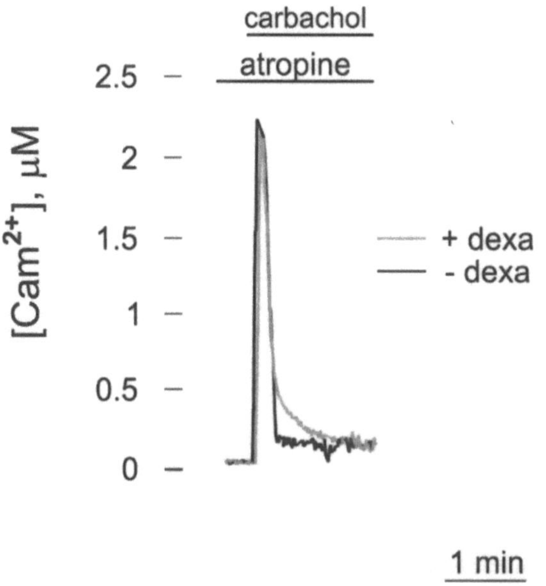

Fig. 7. Monitoring of $[Ca^{2+}]_c$ and $[Ca^{2+}]_n$ in skeletal muscle cells with the nu/cytAEQ chimera. The two compartments were analysed in parallel batches of cells transfected with the nu/cytAEQ plasmid, differing only for dexamethasone pretreatment (±dexa). In the case of $[Ca^{2+}]_n$ measurements, the culture medium was supplemented (24 h before the experiment) with 10 μM dexamethasone, which induced the complete translocation of the recombinant photoprotein to the nucleus. Where indicated the cells were stimulated with carbachol in the presence of atropine.

Fig. 7 shows the monitoring of $[Ca^{2+}]$ changes with nu/cytAEQ chimera, in parallel batches of primary cultures of skeletal muscle measured in the two experimental conditions: either preincubated for 24 h with dexamethasone (thus causing the complete translocation of nu/cytAEQ to the nucleus) or with no hormone pretreatment (thus allowing the retention of the probe in the cytoplasm). The nu/cytAEQ chimera was shown to undergo a hormone-dependent nuclear translocation also in the myotubes (Brini et al., 1997). Upon nicotinic stimulation with carbachol (in presence of atropine) a large rise in $[Ca^{2+}]$ was observed in both compartments, with very similar amplitude and kinetics. The $[Ca^{2+}]$ changes are virtually superimposed and very similar. The mean peak increases in $[Ca^{2+}]$ induced by carbachol were 2.21 ± 0.2 (n=20) and 2.07 ± 0.47 μM (n=11) with and without dexamethasone pretreatment, respec-

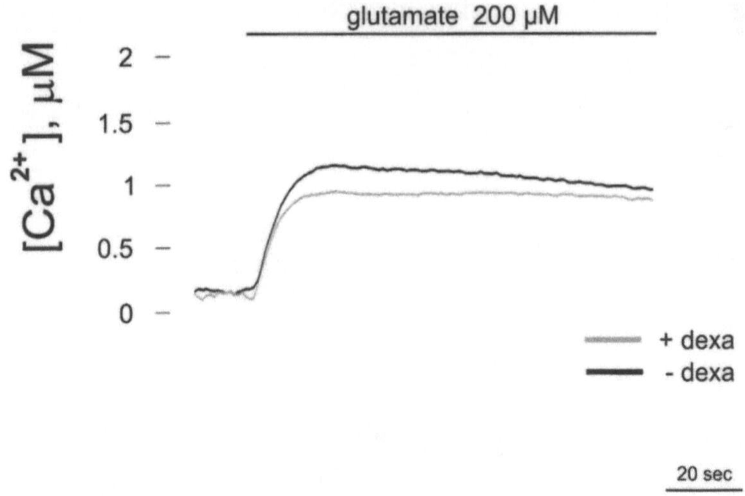

Fig. 8. Monitoring of $[Ca^{2+}]_c$ and $[Ca^{2+}]_n$ in neuronal cells with the nu/cytAEQ chimera. Cells were transfected with nu/cytAEQ chimera. Other conditions as in Fig. 6. Where indicated the cells were stimulated with 200 µM glutamate.

tively. No difference was observed between $[Ca^{2+}]_c$ and $[Ca^{2+}]_n$ when stimulus was either caffeine or KCl (stimuli specific for myotubes, see Brini et al., 1997). Similar results were obtained with the constitutive nuclear aequorin chimera (data not shown).

Fig. 8 shows the monitoring of $[Ca^{2+}]$ changes with nu/cytAEQ chimera in parallel batches of primary cultures of cortical neurons measured after treatment or not with 2 µM dexamethasone for 2 h. The $[Ca^{2+}]_n$ changes closely parallel with the $[Ca^{2+}]_c$ changes both at rest condition and after stimulation with glutamate, confirming previous data obtained in HeLa and muscle cells.

Conclusions

Taken together the above mentioned data suggest that the nuclear membrane does not represent an important barrier to the diffusion of Ca^{2+} from the cytosol to the nucleus and that this latter compartment is in rapid equilibrium with the cytosol.

This main observation initially obtained in HeLa cells, has been confirmed also in highly differentiate cell types such as myotubes and in neurons, thus suggesting that it represents a general concept in intracellular Ca^{2+} homeostasis. However, our data by no means exclude that under some experimental conditions, or in some cell types, nucleus/cytosol gradients can be generated. Indeed, although our data indicate that cytosolic and nucleoplasmic $[Ca^{2+}]$ most often equilibrate, situations may be envisaged in which the two signals may dissociate (e.g. during fertilisation, a large increase in $[Ca^{2+}]_n$ has been hypothesised to occur, Stricker et al., 1992). The possibility of selectively monitoring $[Ca^{2+}]_n$ in these and other models, allows the direct investigation of the mechanism through which the agonist-induced variations of $[Ca^{2+}]_c$ are transduced to the nucleus and modulate processes of large biological and medical relevance, such as gene expression, apoptosis and excitotoxicity.

Acknowledgements. The experimental work described in this chapter was supported by grants from "Telethon" (project n 1285), from the "Human Frontier Science Program", from the "Armenise Foundation" (Harvard), from the Italian University Ministry and from the Italian Space Agency (ASI) to Tullio Pozzan and Rosario Rizzuto.

References

Allen DG, Blinks JR. Calcium transients in aequorin-injected frog cardiac muscle. Nature 1978; 273:509-513.

Barrero, M. J., M. Montero, and J. Alvarez (1997). Dynamics of $[Ca^{2+}]$ in the endoplasmic reticulum and cytoplasm of intact HeLa cells: a comparative study. J. Biol. Chem., 272:27694-27699.

Bastianutto, C., Clementi, E., Codazzi, F., Podini, P., De Giorgi, F., Rizzuto, R., Meldolesi, J., and Pozzan, T. (1995). Overexpression of calreticulin increases the Ca^{2+} capacity of rapidly exchanging Ca^{2+} stores and reveals aspects of their lumenal microenvironment and function. J. Cell. Biol. 130:847-855.

Blinks, J. R., P. H. Mattingly, B. R. Jewell, M. vanLeewen, G. C. Marrer, and D. Allen (1978). Practical aspects of the use of aequorin as a calcium indicator: assay, preparation, microinjection,and interpretation of signals. Meth. Enzymol., 57:292-328.

Breeuwer, M., and Goldfarb, D.S. (1990). Facilitated nuclear transport of histone H1 and other small nucleophilic proteins. Cell 60, 999-1008.

Brini, M., M. Murgia, L. Pasti, D. Picard, T. Pozzan, and R. Rizzuto (1993). Nuclear Ca^{2+} concentration measured with specifically targeted recombinant aequorin. Embo J., 12:4813-4819.

Brini, M., R. Marsault, C. Bastianutto, T. Pozzan, and R. Rizzuto (1994). Nuclear targeting of aequorin. A new approach for measuring Ca^{2+} concentration in intact cells. Cell Calcium, 16:259-268.

Brini, M., R. Marsault, C. Bastianutto, J. Alvarez, T. Pozzan, and R. Rizzuto (1995). Transfected aequorin in the measurement of cytosolic Ca^{2+} concentration ($[Ca^{2+}]_c$): a critical evaluation. J. Biol. Chem., 270:9896-9903.

Brini, M., F. De Giorgi, M. Murgia, R. Marsault, M. L. Massimino, M. Cantini, R. Rizzuto, and T. Pozzan (1997). Subcellular analysis of Ca^{2+} homeostasis in primary cultures of skeletal myotubes. Mol. Cell. Biol., 8:129-143.

Cantini, M., Massimino, ML., Catani, C., Rizzuto, R., Brini, M., and Carraro, U. (1994). Gene transfer into satellite cell from regenerating muscle: bupivacaine allows <greek>b</greek>-Gal transfection and expression in vitro and in vivo. In Vitro Cell. Dev. Biol. Anim., 30A: 131-133.

Collart, M. A., N. Tourkine, D. Belin, P. Vassalli, P. Jeanteur, and J.-M. Blanchard (1991). c-fos gene transcription in murine macrophages is modulated by a calcium-dependent block to elongation in intron 1. Mol. Cell. Biol., 11:2826-2831.

De Robertis, E.M., Longthorne, R.F., and Gurdon, J.B. (1978). Intracellular migration of nuclear proteins in Xenopus oocytes. Nature 272: 254-256.

Dingwall, C., Sharnick, S.V. and Laskey, R.A. (1982). A polypeptide domain that specifies migration of nucleoplasmin into the nucleus. Cell 30: 449-458.

Dingwall, C., and R. A. Laskey (1991). Nuclear targeting sequences-a consensus? Trends Biochem. Sci., 16: 478-481.

Field, J., Nikawa, J., Broek, D., MacDonald, B., Rodgers, L., Wilson, I.A., Lerner, R.A., and Wigler, M. (1988). Purification of a RAS-responsive adenylyl cyclase complex from Saccharomyces cerevisiae by use of an epitope addition method. Mol. Cell. Biol., 8:2159-2165.

Gaido, M. L., and J. A. Cidlowski (1991). Identification, purification and characterization of a calcium-dependent endonuclease (NUC18) from apoptotic rat thymocytes. J. Biol. Chem., 266:18580-18585.

Grynkiewicz, G., M. Poenie, and R. Y. Tsien (1985). A new generation of Ca^{2+} indicators with greatly improved fluorescence properties. J. Biol. Chem., 260:3440-3450.

Hartl, F. U., N. Pfanner, D. W. Nicholson, and W. Neupert (1989). Mitochondrial protein import. Biochim. Biophys. Acta, 988:1-45.

Inouye, S., M. Noguchi, Y. Sakaki, Y. Takagi, T. Miyata, S. Iwanaga, T. Miyata, and F. I. Tsuji (1985). Cloning and sequence analysis of cDNA for the luminescent protein aequorin. Proc. Natl. Acad. Sci. USA, 82:3154-3158.

Kalderon, D., B. L. Roberts, W. D. Richardson, and A. E. Smith (1984). A short amino acid sequence able to specifiy nuclear location. Cell, 39:499-509.

Kendall, J. M., G. Sala-Newby, V. Ghalaut, R. L. Dormer, and A. K. Campbell (1992). Engineering the Ca^{2+}-activated photoprotein aequorin with reduced affinity for calcium. Biochem. Biophys. Res. Comm., 187:1091-1097.

Lievremont, J.P., Rizzuto, R., Hendershot, L., and Meldolesi, J. (1997). BiP, a major chaperone protein of the endoplasmic reticulum lumen, plays a direct and important role in the storage of the rapidly exchanging pool of Ca^{2+}. J. Biol. Chem., 272:30873-30879.

Marsault, R., M. Murgia, T. Pozzan, and R. Rizzuto (1997). Domains of high Ca^{2+} beneath the plasma membrane of living A7r5 cells. Embo J., 16:1575-1581.

Montero, M., M. Brini, R. Marsault, J. Alvarez, R. Sitia, T. Pozzan, and R. Rizzuto (1995). Monitoring dynamic changes in free Ca^{2+} concentration in the endoplasmic reticulum of intact cells. Embo J., 14:5467-5475.

Moreland, R.B., Langevin, G.L., Singer, R.H., Garcea, R.L., and Hereford, L.M. (1987). Amino acid sequences that determine the nuclear localization of yeast histone 2B. Mol. Cell. Biol. 7, 4048-4057.

Nomura M, Inouye S, Ohmiya Y, and Tsuji F.I. (1991) A C-terminal proline is required for bioluminescence of the Ca^{2+}-binding photoprotein, aequorin. FEBS Lett., 295,63-6.

Peters, R. (1986). Fluorescence microphotolysis to measure nucleocytoplasmic transport and intracellular mobility. Biochem. Biophys. Acta, 864, 305-359.

Picard D. and Yamamoto K.R. (1987). Two signals mediate hormone-dependent nuclear localization of the glucocorticoid receptor. Embo J., 6, 3333-3340

Pinton, P., T. Pozzan, and R. Rizzuto (1998). The Golgi apparatus is an inositol 1,4,5 trisphosphate Ca^{2+} store, with distinct functional properties from the endoplasmic reticulum. Embo J., 18:5298-5308.

Ridgway, E. B., and C. C. Ashley (1967). Calcium transients in single muscle fibers. Biochem. Biophys. Res. Comm., 29:229-234.

Ridgway, E. B., J. C. Gilkey, and L. F. Jaffe (1977). Free calcium increases explosively in activating medaka eggs. Proc. Natl. Acad. Sci. USA, 74:623-627.

Rizzuto, R., A. W. M. Simpson, M. Brini, and T. Pozzan (1992). Rapid changes of mitochondrial Ca^{2+} revealed by specifically targeted recombinant aequorin. Nature, 358:325-328.

Rizzuto, R., M. Brini, C. Bastianutto, R. Marsault, and T. Pozzan (1995). Photoprotein mediated measurement of $[Ca^{2+}]$ in mitochondria of living cells. Meth. Enzymol., 260:417-428.

Rizzuto, R., P. Pinton, W. Carrington, F. S. Fay, K. E. Fogarty, L. M. Lifshitz, R. A. Tuft, and T. Pozzan (1998). Close contacts with the endoplasmic reticulum as determinants of mitochondrial Ca^{2+} responses. Science, 280:1763-1766.

Rusconi S, and Yamamoto K.R. (1987) Functional dissection of the hormone and DNA binding activities of the glucocorticoid receptor. Embo J., 6,1309-1315

Rutter, G. A., P. Burnett, R. Rizzuto, M. Brini, M. Murgia, T. Pozzan, J. M. Tavare, and R. M. Denton (1996). Subcellular imaging of intramitochondrial Ca^{2+} with recombinant targeted aequorin. Significance for the regulation of pyruvate dehydrogenase activity. Proc. Natl. Acad. Sci. USA, 93:5489-5494.

Shimomura, O., Musicki, B., and Kishi, Y. (1989) Semi-syntetic aequorins with improved sensitivity to Ca^{2+} ions. Biochem. J. 261:913-920.

Stricker, S.A., Centonze, V.E., Paddock, S.W. and Scatten, G. (1992) Confocal microscopy of fertilization-induced calcium dynamics in sea urchin eggs. Dev. Biol. 149:370-380.

Tombes, R. M., C. Simerly, G. G. Borisy, and G. Schatten (1992). Meiosis, egg activation, and nuclear envelope breakdown are differentially reliant on Ca^{2+}, whereas germinal vesicle breakdown is Ca^{2+} independent in the mouse oocyte. J. Cell Biol., 117:799-811.

Tsien, R. Y., T. Pozzan, and T. J. Rink (1982). T-cell mitogens cause early changes in cytoplasmic free Ca^{2+} and membrane potential in lymphocytes. Nature, 295:68-71.

Zhao, L., and Padmanabhan, R. (1988). Nuclear transport of adenovirus DNA polymerase is facilitated by interaction with preterminal protein. Cell, 55, 1005-1015.

Measuring Ca^{2+} in Isolated Nuclei

OLEG GERASIMENKO

Introduction

Ca^{2+} is important for the control of fundamental nuclear processes such as gene expression, DNA replication and repair, phosphorylation of nuclear proteins, and apoptosis (Tombes 1992). It has been shown recently that fusion of nuclear vesicles (Sullivan 1993) and cell cycle regulation (Means 1994) require Ca^{2+} mobilization. It is known that Ca^{2+} spikes play a complex role in completion of meiosis and initiation of mitosis in marine eggs (Means 1994). The nucleus accumulates Ca^{2+} via endoplasmic reticulum type ATP-dependent Ca^{2+} pump (Lanini 1992). The nuclei contain Ca^{2+} binding protein calmodulin and the main binding protein of the endoplasmic reticulum, calreticulin (Serratosa 1988; Burns 1994). The functional receptors for the Ca^{2+}-releasing messenger inositol 1,4,5-trisphosphate (Nicotera 1990; Malviya 1990; Kume 1993), and mechanism for production of this messenger (Divecha 1993) are also present in the nucleus.

The nuclear envelope of higher eukariotic cells is formed from two concentric membranes. The outer nuclear membrane is continuous with the endoplasmic reticulum (ER) membrane and has ER properties, for example protein composition which is different from the inner membrane (Newport 1987). The perinuclear space (space between the inner and outer nuclear membrane) is in direct continuity with the ER lumen (Gerasimenko 1996). Nuclear pores are assembled by the nuclear pore complex

Oleg Gerasimenko, The University of Liverpool, The Physiological Laboratory, Crown Street, P.O. Box 147, Liverpool, L69 3BX, UK (*phone* +44-151-794-5357; *fax* +44-151-794-5327; *e-mail* olegl@liverpool.ac.uk)

(NPC). The large supramolecular assemblies (approximately 125 MDa) contain more than 100 different protein components called nucleoporins mediating the traffic between the nucleoplasm and cytosol (Sweet 1995). The macromolecules such as most proteins and ribonucleic acids (RNAs) are transported by the temperature and energy-dependent process that requires activity of specific NPC proteins with nuclear localisation sequence as well as cytosolic transport factor (Sweet 1995). Small molecules and ions with molecular mass up to 10–20 kDa can easy diffuse through the NPC at physiological concentrations.

On the other hand differences between Ca^{2+} concentration in the nucleoplasm and the cytosol have been observed, particularly after stimulation (Berridge 1993). However, it has been shown that the previously reported amplification of changes in free Ca^{2+} concentration in the nucleus is a measurement artefact (Gillot 1993, Al-Mohanna 1994). In 1993 Brini et al measured Ca^{2+} concentration in nucleus using a specifically targeted recombinant aequorin and these data indicate that, in general, cytosolic and nucleoplasmic Ca^{2+} concentrations equilibrate rapidly.

On the basis of work on isolated liver nuclei, it has been concluded that there is ATP - dependent Ca^{2+} uptake into and IP_3 evoked release of Ca^{2+} from the intranuclear space (Nicotera 1989, 1990). The most direct measurements of the nucleoplasmic Ca^{2+} concentration in intact cells (Brini 1993) indicate that the nuclear envelope is very permeable to Ca^{2+}. It is difficult to understand how the mechanisms previously proposed for nuclear Ca^{2+} homeostasis, based on work on isolated nuclei (Nicotera 1989,1990, 1992), can operate. In 1995 Gerasimenko et al. investigated Ca^{2+} transport and the fluorescent Ca^{2+}-sensitive dye distributions in isolated liver nuclei using digital imaging and confocal microscopy. These data show that when isolated nuclei are exposed to the cell-permeable fluorescent dye fura-2 acetoxymethylester (fura-2 AM), the fura-2 indicator is concentrated in or around the nuclear envelope with a very low concentration in the nucleoplasm. The distribution is virtually identical to that of the ER marker 3,3'-dihexyloxacarbocyanine iodide ($DiOC_6(3)$). It is therefore clear that the Ca^{2+} accumulation measured is due to uptake into a peripheral envelope of stores with the ER properties and not into the nucleoplasm. It has been shown that when the nucleoplasmic Ca^{2+} concentration is measured using dextrans attached to fluorescent calcium-sensitive dyes, changes in the external Ca^{2+} concentration produce similar rapid changes in the nucleoplasm, indicating a high degree of permeability across the nuclear envelope. Both the Ca^{2+} releasing messengers IP_3 and cyclic ADP-ribose evoke release of accumulated Ca^{2+} directed into the nucleoplasm from where it can diffuse out through the nuclear pores (nuclear pore complex).

Outline

The measurements of the free Ca^{2+} uptake and release from the isolated liver nuclei using Ca^{2+}-sensitive fluorescent probes with the help of digital imaging and confocal microscopy.

Materials

- Laser scanning confocal microscope (distribution studies and Calcium Green measurements)
- Imaging system (fura (Mag-fura) measurements)
- Homogeniser (glass/Teflon)
- Cheesecloth
- Centrifuge (for 1,000 x g)
- Fura-2 pentapotassium salt, fura-2 AM, Mag fura-2 AM, fura-2 dextran (MW 70,000), Calcium Green-1 AM, Calcium Green-1 dextran (MW 70,000), ethidium bromide, $DIOC_6(3)$ (Molecular Probes)
- Inositol 1,4,5,-trisphosphate (IP_3), ionomycin, adenosin trisphosphate (ATP), thapsigargin (Sigma)
- cyclic adenosin diphosphate-ribose (cADPr) (Amersham)
- ryanodine (Calbiochem)

Bath solution	Comments
125 mM KCl	
2 mM K_2HPO_4	
50 mM HEPES	
4 mM $MgCl_2$	
0.1 mM EGTA	
1 mM ATP	
1 mg/ml trypsin inhibitor (from soybean type I-S)	for isolation procedure only
0.25 mM sucrose	for isolation procedure only
adjust to pH 7.0	

Procedure

Preparation of isolated mouse liver nuclei

1. Perform all steps at 4 °C. Wash the liver from male CD1 mouse in the K Hepes bath solution supplemented with 0.25 mM sucrose and 1 mg/ml trypsin inhibitor, then cut into small pieces with scissors and homogenize with glass/Teflon homogenizer (8-10 strokes) in the same buffer.
2. Filter the homogenate through several layers of cheesecloth.
3. Centrifuge the homogenate 1,000 x g for 1 minute and resuspend the pellet of the nuclei in the K Hepes buffer with 0.25 mM sucrose and 1 mg/ml trypsin inhibitor.
4. Homogenize the suspension of the nuclei with glass/Teflon homogenizer (5 strokes).
5. Centrifuge the homogenate 1,000 x g for 1 minute and gently resuspend the pellet of the nuclei in the K Hepes buffer.

Fluorescent labelling procedures

1. Nuclear envelope measurements.
Incubate the nuclei with K Hepes bath solution containing 20 μM of fura-2 AM or Calcium Green-1 AM or Mag-fura-2 AM for 45 minutes at 4 °C. Wash loaded nuclei by centrifugation 1,000 x g for 1 minute.

2. Nucleoplasm measurements.
   Alternatively load nuclei with 20 μM of fura-2 dextran or Calcium Green-1 dextran by incubation for 10 minutes at 4 °C in the standard bath solution. These fluorescent dextrans accumulate and remain in nuclei for the 30 minutes. Washing procedure by centrifugation 1,000 x g for 1 minute is necessary to reduce the background fluorescence.

3. Additional staining.
   Stain the nuclei with ethidium bromide in concentration 5 μg/ml or $DiOC_6$ in concentration 2.5 μg/ml for 5 minutes at room temperature and wash once by centrifugation 1,000 x g 1 minute with K Hepes bath solution.

Fluorescent measurements

1. Place sample of nuclei loaded with the fluorescent indicator in the perfusion experimental chamber (room temperature) and wash them by perfusion with standard solution for 10-15 minutes before the beginning of the experiment. This step is crucial for complete cleavage of acetoxy-methyl (AM) ethers of AM forms of fluorescent dyes.

2. Use staining with ethidium bromide (exitation is 360 nm, emission > 400 nm) at the end of the experiment for additional nuclear control.

3. Distribution of the fluorescence of the fura-2 (loaded in AM form) or fura-2 dextran in single isolated nuclei could be obtained using any imaging system capable of using excitation 360 nm (the isobestic point of fura-2) and emission 510 nm, although for best results UV excitation confocal microscope should be used.

4. Fluorescence of $DiOC_6(3)$-loaded nuclei can be measured with excitation 488 nm and emission > 515 nm.

Results

Distribution of fluorescent indicators in isolated nuclei

Measurements of fluorescence in the experiments with isolated nuclei loaded with fura-2 in AM form were done using the excitation wavelength of 360 nm (the isobestic point of fura-2 is 360 nm) by the imaging system and confocal microscopy (Fig. 1). The intensity of the fluorescence is relatively high in

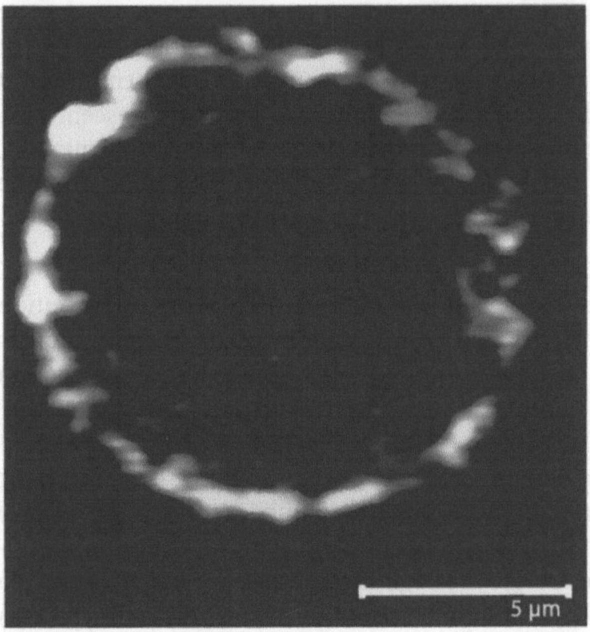

Fig. 1. Distribution of calcium-sensitive probe fura-2 in single isolated liver nucleus, measured with a confocal system (excitation 363 nm, emission >515 nm). Bar corresponds to 5 μM. Note ring distribution of fluorescence at the perifery of the nucleus. Modified from Gerasimenko 1995.

the perifery of the nucleus but not in the central area. Using the ER marker $DiOC_6(3)$ we obtained similar to fura-2 distribution of the fluorescence in the nucleus. This similarity in fluorescence distribution indicates that fura-2 is localised in the areas of the nuclear envelope and ER, which remain connected to the nuclei after the isolation procedure. No fluorescence was observed in the nucleoplasm region. This ring-like distribution of fura-2 and $DiOC_6(3)$ fluorescence was verified using confocal microscopy (Fig. 1, 2).

The intensity of the fluorescence of fura-2 dextran in the nuclei was highest in the central area and lowest in the periphery (Fig. 3). The same type of distribution was obtained for Calcium Green-1 dextran. The distribution of the fluorescence of the DNA-RNA marker ethidium bromide is similar to the distribution of the fluorescence of the dextrans labelled with fluorescent Ca^{2+} indicators. This bell-shaped distribution of the fluorescent dextrans indicates that these molecules accumulate in the nucleoplasm.

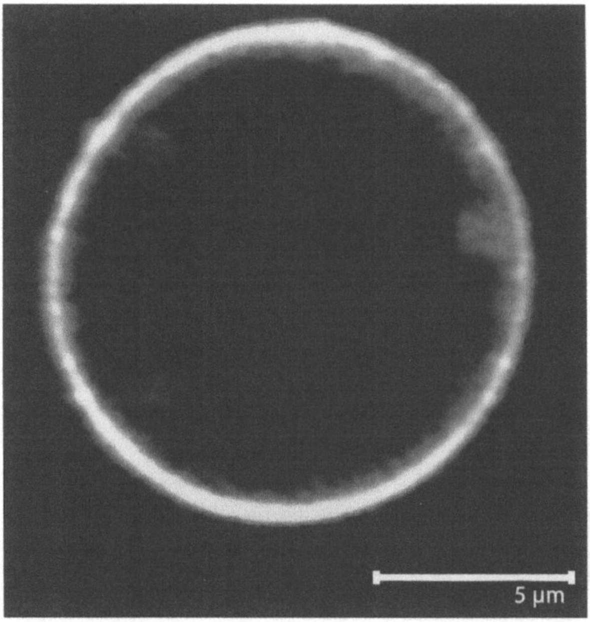

Fig. 2. Distribution of endoplasmic reticulum marker DiOC6(3) in isolated liver nucleus measured with a confocal system (excitation 488 nm, emission > 515 nm). Bar corresponds to 5 µM. Note ring distribution of fluorescence. Modified from Gerasimenko 1995.

Ca^{2+} transport into and out of isolated single nuclei

Changes of the external Ca^{2+} concentration (10^{-8} M to 1 mM) did not affect the fluorescence of fura-2-loaded (in AM form) nuclei. However, addition of ionomycin to an external solution with a high Ca^{2+} concentration (1 mM) resulted in saturation of the fura-2 ratio signal, and a subsequent wash with the Ca^{2+} chelator EGTA in the presence of ionomycin induced recovery of the signal followed by a continuous slow decline of the nuclear Ca^{2+} concentration (Fig. 4). Figure 5 represents an ATP-dependent accumulation of Ca^{2+} into the nuclei. After addition of 1 mM ATP free Ca^{2+} increases during next 200 s from 100-200 nM to 500-1000 nM (Gerasimenko 1995). The calcium sensitive dye fura-2 has a range of measurements optimised for cytoplasmic Ca^{2+} changes. It is known that the nuclear envelope is connected to the endoplasmic reticulum where free Ca^{2+} concentrations are in the range of hundreds of micromols. For endoplasmic reticulum Ca^{2+} measurements the low affinity calcium sensitive dyes are usually used. The similar experiments were done with the nuclei loaded with Mag-fura-2 AM (Ca^{2+} Kd -

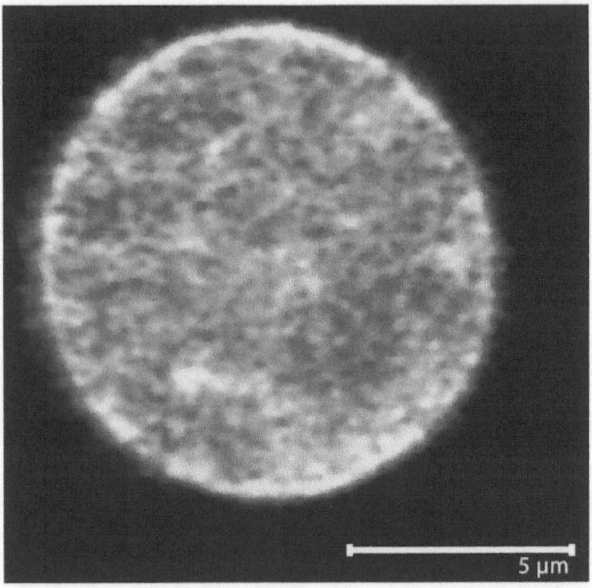

Fig. 3. Distribution of the dextran-linked fura-2 in single isolated liver nucleus, measured with a confocal system. Bar corresponds to 5 µM. Note that fluorescence is uniformly disributed throughout the nucleoplasm. Modified from Gerasimenko 1995.

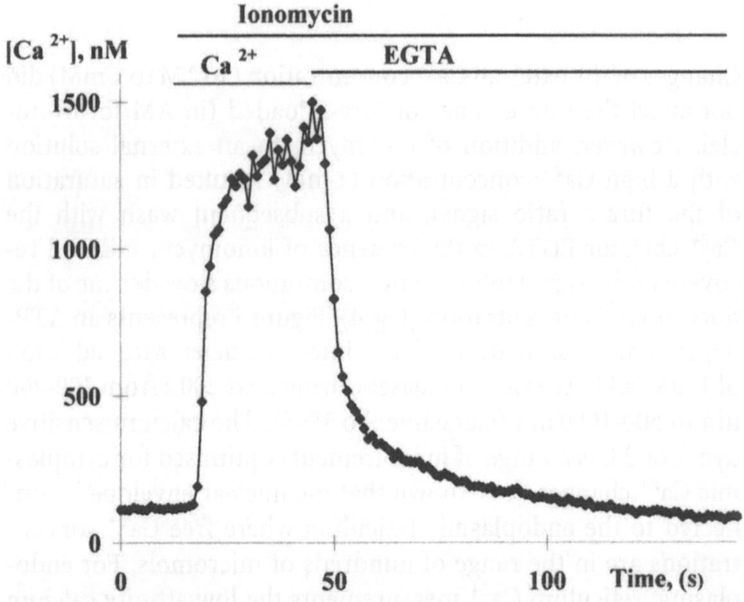

Fig. 4. Subsequent application of high Ca^{2+} (1mM) and EGTA (2 mM) to a single isolated nucleus in the presence of 20 µM ionomycin. Nuclei were loaded with fura-2 AM. Modified from Gerasimenko 1995.

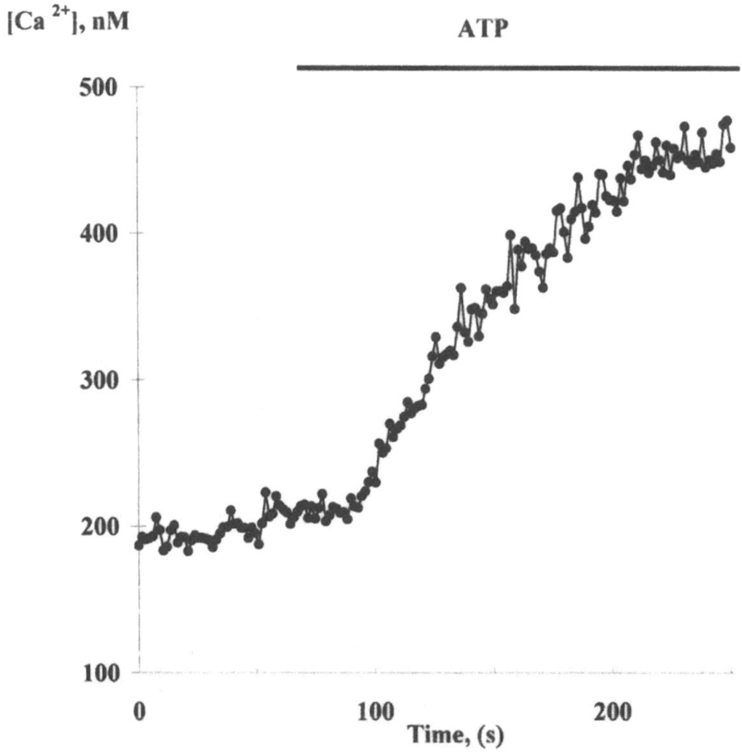

Fig. 5. ATP-dependent Ca^{2+} uptake in the nuclear envelope. Application of 1 mM ATP to a nucleus in standard solution containing 100 nM free Ca^{2+}. Nuclei loaded with fura-2 AM. Modified from Gerasimenko 1995.

25 μM, Molecular Probes) (Fig. 6). Calcium increases after addition of ATP for 200 s and remains at the saturated level, but in Mag-fura-2 loaded nuclei the saturation level was in the range of 150 ± 50 μM of free Ca^{2+}. ATP dependent Ca^{2+} accumulation in nuclei was inhibited by 1μM of thapsigargin.

Accumulated Ca^{2+} could be released by IP_3, but this effect was difficult to detect in the ATP-containing solution. After removal of ATP from the external buffer, IP_3 induces significant decrease in the nuclear Ca^{2+} concentration (Fig. 7). These results are in agreement with the experiments using suspensions of isolated nuclei (Nicotera 1989, 1990, Hechtenberg 1993). However, the experiments with single nuclei show that ATP dependent Ca^{2+} accumulation and IP_3 induced release occur into and from stores around the nuclear envelope.

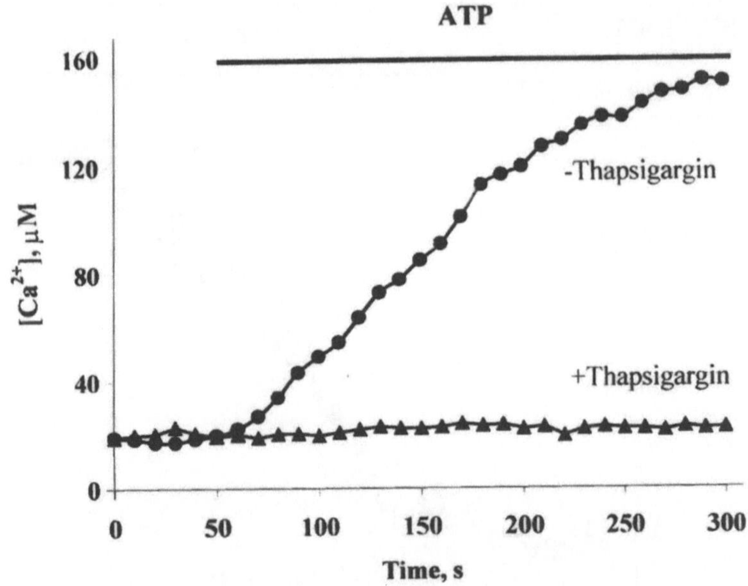

Fig. 6. ATP-dependent Ca^{2+} uptake in the nuclear envelope using Mag-fura-2 AM loaded nuclei. Application of 1 mM ATP to a nucleus in standard solution containing 100 nM free Ca^{2+}. No calcium uptake has been observed in the presence of thapsigargin (lower trace).

Since there is growing evidence for a Ca^{2+} releasing messanger role for cADPr in many systems (Lee 1989, Hua 1994, Thorn 1994) the effect of this agent has been investigated. In our experiments 10 µM cADPr evoked a clear loss of Ca^{2+} from the nuclear envelope. The accumulation of dextrans labelled with Ca^{2+}-sensitive indicator in the internal areas of the nuclei (see Fig. 3) gave us a unique opportunity to monitor Ca^{2+} changes in the nucleoplasm. Changes of the Ca^{2+} concentration in the external solution resulted in rapid changes of the fluorescence of the dextran-bound Ca^{2+} indicators.

An increase of the external Ca^{2+} concentration to 1 mM caused saturation of the indicators in the nucleoplasm, whereas a subsequent application of EGTA (2 mM) resulted in reversal of the signal, indicating that Ca^{2+} was being removed from the nucleoplasm (Fig. 8). These rapid changes of fluorescence of the dextran-bound indicators strongly suggest that the nuclear envelope is easily permeable to Ca^{2+}. However, stimulation of nu-

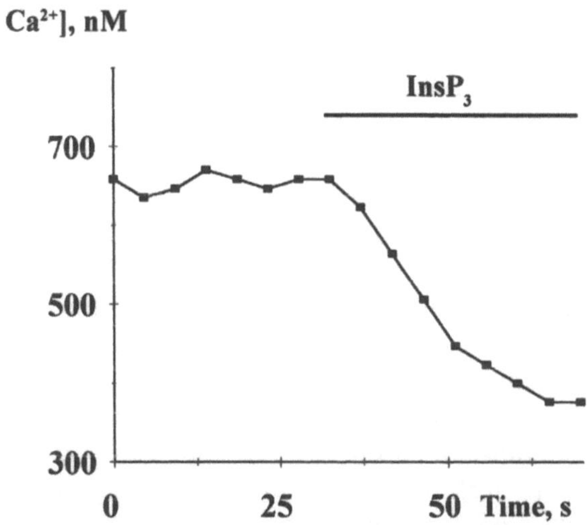

Fig. 7. IP_3 evoked Ca^{2+} release from nuclear envelope. Nuclei were incubated with 200 nM free Ca^{2+} in standard solution in the presence of 1 mM ATP for 2 min were washed with standard solution without Ca^{2+} and ATP, and 10 μM IP_3 was applied. Modified from Gerasimenko 1995.

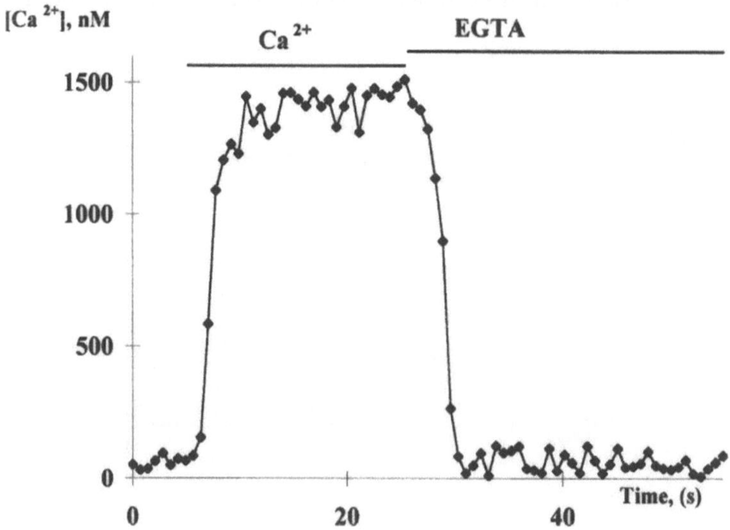

Fig. 8. Calcium ions can move easily across the nuclear envelope. Subsequent applications of high Ca^{2+} (1 mM) and EGTA (2 mM) to a single isolated nuclei. Nuclei loaded with fura-2 dextran. Modified from Gerasimenko 1995.

clei that accumulated Ca^{2+} in the presence of ATP with IP_3 (Fig. 9) and cADPr induced a rapid transient elevation of the intranuclear Ca^{2+} concentration measured with the dextran-bound Ca^{2+} indicators (Gerasimenko 1995). The effect of cADPr indicates the presence of functional ryanodine receptors (Meszaros 1993), and we therefore also tested the actions of ryanodine. Ryanodine itself evoked a small but clear rise in the nucleoplasmic Ca^{2+} concentration (Gerasimenko 1995). This effect was transient but with a rather longer duration than those induced by IP_3 and cADPr. The dextran-bound Ca^{2+} indicators that accumulate inside the nucleus appear to record Ca^{2+} release from the nucear envelope, and this release is directed into the nucleoplasm. The transient nature of these responses (Fig. 8, 9) is probably due to leakage of Ca^{2+} from the nucleoplasm through the nuclear pore complexes (Gerasimenko 1995).

Figure 10 summarizes previous conclusions (Nicotera 1989, 1990, 1992) as well as the main conclusions from the results presented in (Gerasimenko 1995, 1996) concerning Ca^{2+} transport into and out of isolated nuclei.

The principal results of these studies are:

- The distribution of fluorescence of calcium sensitive indicator fura-2 is non homogeneous but concentrates in the nu-

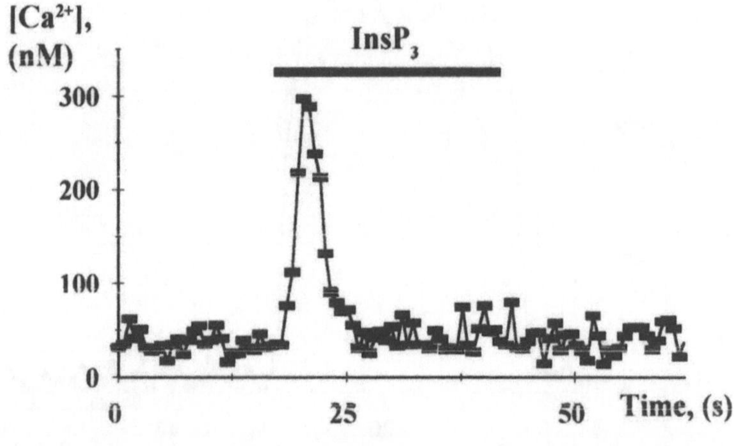

Fig. 9. IP_3 cause Ca^{2+} release directed into the nucleoplasmic space. Nucleus loaded with Ca^{2+} in the presence of ATP was washed with standard solution (without Ca^{2+} and ATP), and 10 μM IP_3 was applied. Nuclei loaded with fura-2 dextran. Modified from Gerasimenko 1995-

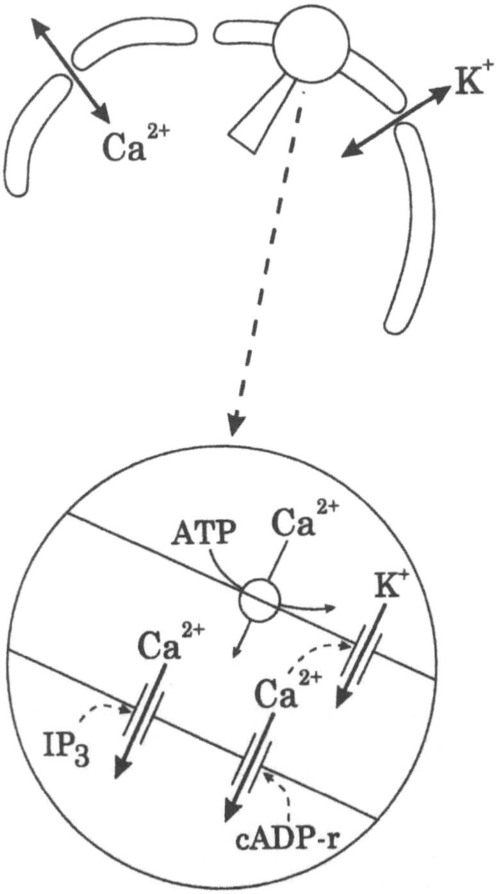

Fig. 10. Model of nuclear Ca^{2+} transport. Modified from Gerasimenko 1995.

clear envelope. The fluorescence of ER marker DiOC6(3) has the same distribution.

- The fluorescent dextran-bound calcium indicators fura-2 dextran and Calcium Green-1 dextran easily go through the nuclear pore complex and concentrate into nucleoplasm.
- The ATP-dependent Ca^{2+} uptake into isolated nuclei occurs into the nuclear envelope. Ca^{2+} transport does not occur into and out of the nucleoplasmic space, but into and out of the nuclear envelope.

- The application of IP_3 to single isolated nuclei evokes a release of the accumulated Ca^{2+} from the nuclear envelope. The application of cADP-ribose to single isolated nuclei induces a release of the accumulated Ca^{2+} from the nuclear envelope. These results directly indicate the presence of functional receptors in the nuclear envelope.
- The functional IP_3 or cADP-ribose evoked calcium release from nuclear envelope directed into the nucleoplasm.
- The functional IP_3 or cADP-ribose evoked calcium release from nuclear envelope directed into the nucleoplasm presumably results in a specific nuclear calcium signalling system that is of great importance for many intranuclear processes.

Acknowledgements. This work was supported by an MRC Programme Grant (G 8801575).

References

Al-Mohanna FA, Caddy KWT, Bolsover SR (1994) The nucleus is insulated from large cytosolic calcium ion changes. Nature 367:745-750

Berridge MJ (1993) Inositol trisphosphate and calcium signalling. Nature 361:315-325

Brini M, Murgia M, Pasti L, Picard D, Pozzan T, Rizzuto R (1993) Nuclear Ca^{2+} concentration measured with specifically targeted recombinant aequorin. EMBO Journal 12:4813-4819

Burns K, Duggan B, Atkinson EA, Famulski KS, Nemer M, Bleackley RC, Michalak M (1994) Nature 367:476-480

Divecha N, Banfic H, Irvine RF (1993) Inositides and the nucleus and inositides in the nucleus. Cell 74:405-407

Gerasimenko O, Gerasimenko J, Tepikin A, Petersen O (1995) ATP-dependent accumulation and inositol trisphosphate- or cyclic ADP-ribose-mediated release of Ca^{2+} from the nuclear envelope. Cell 80:439-444

Gerasimenko O, Gerasimenko J, Tepikin A, Petersen O (1996) Calcium transport pathway in the nucleus. Pflügers Archiv 432:1–6

Gillot I, Whitaker M (1993) Imaging calcium waves in eggs and embryos. Journal of Experimental Biology 184:213-219

Kume S, Muto A, Aruya J, Nakagawa T, Michikawa T, Furuiche T, Nakade S, Okano H, Mikoshiba K (1993) The Xenopus IP_3 receptor: structure, function, and localisation in oocytas and eggs Cell 73:669-677

Lanini L, Bachs O, Carafoli E (1992) The calcium pump of the liver nuclear membrane is identical to that of endoplasmic reticulum. Journal of Biological Chemistry 267:11548-11552

Malviya AN, Rogue P, Vincendon G (1990) Stereospecific inositol 1,4,5-[32P] trisphosphate binding to isolated rat liver nuclei: evidence for inositol trisphosphate receptor-mediated calcium release from the nucleus. Proc. Natl. Acad. Sci. USA 87:9270-9274

Means AR (1994) Calcium, calmodulin and cell cycle regulation. FEBS Letters 347:1-4 Newport J, Forbes DJ (1987) The nucleus: structure, function and dynamics. Annu Rev Biochem 56:535-565

Nicotera P, Orrenius S, Nilsson T, Berggren P-O (1990) An inositol 1,4,5-trisphosphate-sensitive Ca^{2+} pool in liver nuclei. Proc. Natl. Acad. Sci. USA 87:6858-6862

Serratosa J, Pujol MJ, Bachs O, Carafoli E (1988) Biochem.Biophys. Res. Commun. 142:1162-1169

Sullivan KMC, Busa WB, Wilson KL (1993) Calcium mobilization is required for nuclear vesicle fusion in vitro: implications for membrane traffic and IP_3 receptor function. Cell 73:1411-1422

Sweet DJ, Gerace L (1995) Taking from the cytoplasm and giving to the pore: soluble transport factors in nuclear protein import. Trends Cell Biol. 5::444-447

Tombes RM, Simerly C, Borisy GG, Schatten G (1992) Meiosis, egg activation and nuclear breakdown are differentially reliant on Ca^{2+}, whereas germinal vesicle breakdown is Ca^{2+}-independent in the mouse oocyte. Journal of Cellular Biology 117:799-811.

Ca^{2+} Measurement with Luminescent Probes in the Endoplasmic Reticulum

JAVIER ALVAREZ and MAYTE MONTERO

Introduction

The photoprotein aequorin can be used to measure specifically the $[Ca^{2+}]$ inside the lumen of the endoplasmic reticulum ($[Ca^{2+}]_{ER}$) of intact cells. The main advantage of this technique is the highly precise targeting of the probe. The main disadvantage relies in the relatively low amount of emitted light, which makes it difficult to perform single-cell imaging studies at the required time resolution. On the technical side, the equipment required for luminescence measurements in cell populations is simple and low-cost, but expression of ER-targeted aequorin requires previous transfection or infection to introduce the appropriate DNA plasmid, or alternatively the use of stable cell clones. Finally, calibration of the luminescent records to obtain $[Ca^{2+}]$ values is relatively easy, but interpretation of the results requires taking into account that aequorin is being rapidly consumed during the experiment due to the high levels of Ca^{2+} (hundreds of micromolar) present in the ER.

✉ Javier Alvarez, Facultad de Medicina, Universidad de Valladolid y Consejo Superior de Investigaciones Científicas, Departamento de Bioquímica y Biología Molecular y Fisiología, Instituto de Biología y Genética Molecular (IBGM), Ramon y Cajal 7, Valladolid, 47005, Spain (*phone* +34-983-423085; *fax* +34-983-423588; *e-mail* jalvarez@ibgm.uva.es)
Mayte Montero, Facultad de Medicina, Universidad de Valladolid y Consejo Superior de Investigaciones Científicas, Departamento de Bioquímica y Biología Molecular y Fisiología, Instituto de Biología y Genética Molecular (IBGM), Ramon y Cajal 7, Valladolid, 47005, Spain

Principles of measurement

Measuring of $[Ca^{2+}]_{ER}$ with aequorin requires taking into account three basic ideas:

- Targeting and retention of the photoprotein inside the endoplasmic reticulum. This was achieved by fusing the cDNA of aequorin to a portion of the DNA of Igγ2b heavy chain including the leader sequence (L), the VDJ and the CH1 domains (Montero et al., 1995, see Fig.1). The leader sequence signals the protein to traverse the ER membrane, and the CH1 domain is responsible for retention inside the ER via binding to the ER-resident chaperone BiP. In cells producing antibodies, this interaction is displaced by the Ig light chain, leading to secretion. In other cells the chimeric aequorin containing the CH1 domain of the Ig heavy chain is expected to be selectively retained in the ER. As shown in the figure, the construct contains also an epitope tag (HA1) that allows detection by immunofluorescence. Both immunofluorescence and electron microscopy have shown that targeting is extremely selective for the ER (Montero et al., 1995, 1997b).
- Reduction of the Ca^{2+}-affinity of aequorin is essential. Aequorin emits light in a Ca^{2+}-dependent manner when reconstituted with its prosthetic group, coelenterazine. Emission of light produces apoaequorin and the release of the cofactor as coelenteramide. As aequorin reconstitution is relatively slow, emission of light can be considered as essentially irreversible from a practical point of view. Therefore, the amount of active aequorin decreases continuously along the experiment, a phenomenon usually referred to as aequorin consumption. Native aequorin is able to measure $[Ca^{2+}]$ in a range from 0.2 to 5-10 μM. The rate of aequorin

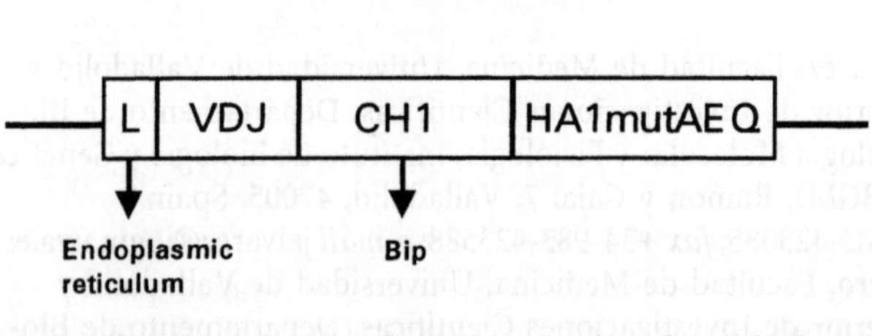

Fig. 1. Chimeric aequorin plasmid targeted to the endoplasmic reticulum.

consumption in the presence of higher [Ca^{2+}] is such that all the aequorin would be consumed within a few seconds. Therefore, to measure $[Ca^{2+}]_{ER}$ values of hundreds of micromolar we must decrease the Ca^{2+}-affinity of aequorin at least by two orders of magnitude. This reduction involves the combination of two steps. A single-point mutation in one of the Ca^{2+}-binding sites (Asp^{119} to Ala) reduces the Ca^{2+}-affinity by about one order of magnitude (Montero et al., 1995). Then, substitution of native coelenterazine by a synthetic derivative that reduces the rate of luminescence emission, coelenterazine n (Shimomura et al., 1993), decreases further the affinity and extends finally the measuring range up to near the millimolar [Ca^{2+}] level (Montero et al., 1997; Barrero et al., 1997).

- Ca^{2+}-depletion of the ER is necessary to allow reconstitution of the photoprotein. Reducing the Ca^{2+}-affinity of aequorin allows measuring $[Ca^{2+}]_{ER}$ levels, but even mutated low-Ca^{2+}-affinity aequorin can hardly be reconstituted with coelenterazine n in the presence of such high $[Ca^{2+}]_{ER}$ levels. This is because the rate of reconstitution is quite slow compared to the rate of light emission at the $[Ca^{2+}]_{ER}$ level of intact cells.

Outline

The protocol of cell preparation for a typical experiment is outlined in Fig. 2.

Materials

Reagents and solutions

- The chimeric ER-targeted mutated aequorin (ERmutAEQ), inserted in the mammalian expression vector pcDNAI, is commercially available from Molecular Probes.
- Transfection or infection methods to express the ERmutAEQ gene in the desired cell types.
- Extracellular medium: NaCl, 145 mM; KCl, 5 mM; $MgCl_2$, 1 mM; glucose, 10 mM; HEPES, 10 mM, pH 7.4

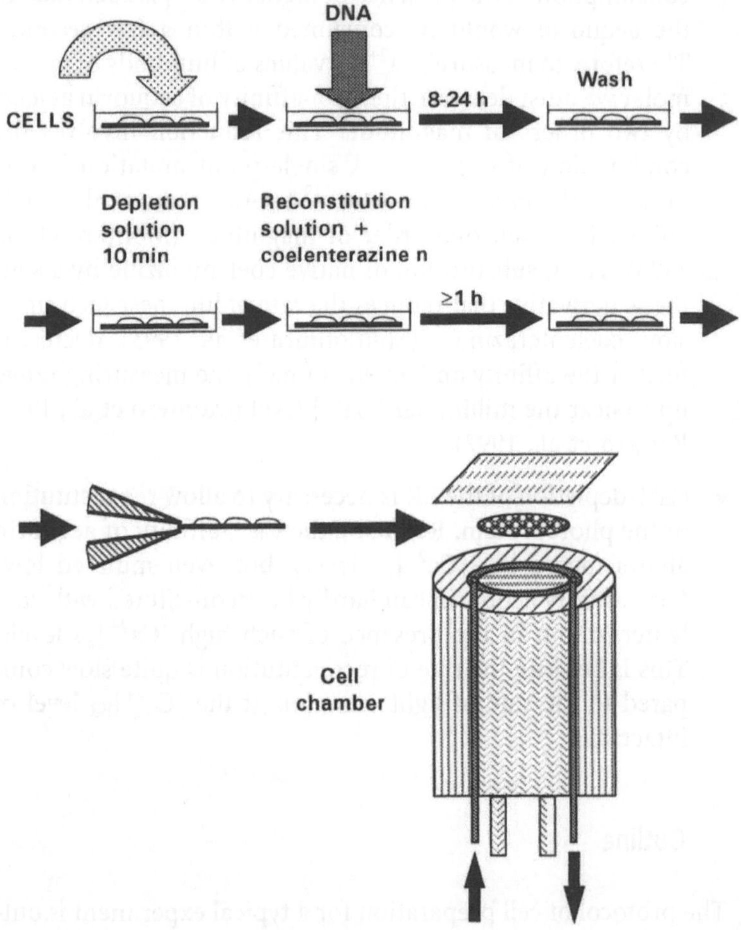

Fig. 2. Outline of a typical experiment.

- Depletion solution: Extracellular medium supplemented with EGTA 3 mM and 2,5-di-tert-butyl-benzohydroquinone (or 2,5-di-tert-butyl-hydroquinone, BHQ) 10 µM. Stocks of BHQ are prepared at 20-50 mM in DMSO and distributed in small aliquots, as frequent freezing and thawing results in loss of activity.
- Reconstitution solution: Extracellular medium supplemented with 0.5 mM EGTA and BHQ 10 µM

- Coelenterazine n (Molecular Probes). Stocks of coelenterazine n are prepared in methanol at 200 μM, and long-term stored in aliquots at -80 °C in nitrogen atmosphere and light protected. The aliquot in use may just be stored at -20 °C and light protected without significant loss in activity in 1-2 month periods.
- Lysis solution: $CaCl_2$ 10 mM; digitonin 100 μM
- Intracellular medium (for experiments with permeabilized cells): NaCl, 10 mM; KCl, 140 mM; $MgCl_2$, 1 mM; KH_2PO_4, 1 mM; ATP-Mg, 2 mM; Hepes, 20 mM, pH 7.0

Equipment

The equipment required for luminescence measurements in cell populations was mentioned earlier in the book. Briefly, we use a Thorn EMI photomultiplier 9789A refrigerated at 4 °C inside a cool box (a small commercial refrigerator may do the job). The 10 mm photocathode is placed in almost direct contact with the thermostatized perfusion chamber (Fig. 2), and covers most of the surface of the glass coverslip. A system of 8 electrovalves allows perfusing different solutions, placed in a nearby bath, during the experiment. The photomultiplier is usually set at 1-1.2 kV, and the output goes via an amplifier-discriminator AD2 up μp to a counter/timer PCB CT1 card in the computer, both from Thorn EMI. Luminescence data are stored every 50 ms, averaged every second and transformed into [Ca^{2+}] values through a home-made software (see below).

Procedure

1. Cells should be seeded at semiconfluency onto glass coverslips. Attachment is required to allow perfusion of solutions during the experiment. To facilitate attachment in some cell kinds, we previously treat the coverslips with poly-L-lysine. Autoclaved coverslips are introduced in a filtered sterile 0.01 mg/ml poly-L-lysine solution for 5 min, washed 2 min in sterile water and finally air dried completely under sterile conditions.
2. Expression of ERmutAEQ. We have stable HeLa cell clones that express ERmutAEQ (Montero et al., 1995). To measure

$[Ca^{2+}]_{ER}$ in other cell types, the ERmutAEQ plasmid should be expressed in the desired cell culture. Description of the different transfection procedures is outside the scope of this chapter. We have used successfully Ca-phosphate and transfection reagents such as Lipofectamine (Gibco) and Fugene (Roche) to express this plasmid in cell lines such as HeLa and CHO. More recently, we have also used herpes virus-based methods to express the gene in several primary cultures, e.g. chromaffin cells, cerebellar granule cells and adenohypophisary cells, as well as in cell lines such as HeLa, GH3 or PC12 (Alonso et al., 1997 and 1999). Adequate expression is usually obtained after 8-12 h using virus-based methods, and 12-24 h using transfection methods.

3. Ca^{2+}-depletion of the ER. The glass coverslip containing the cells that express ERmutAEQ is transferred to a well (of a 24-well plate) containing 1ml of extracellular medium. After washing once with extracellular medium, 0.4 ml of depletion solution is added to the well and left for 10 min at room temperature. This time is enough to empty almost completely the ER of Ca^{2+} in the presence of BHQ and EGTA. BHQ is a reversible inhibitor of the ER-Ca^{2+} pumps (Sarcoplasmic and Endoplasmic Reticulum Ca^{2+} ATPases, SERCA), and produces a rapid Ca^{2+} depletion of the ER (half-time 1–2 min) through some still unknown leak pathways. Ca^{2+} released from the ER is extruded out of the cell by the plasma membrane Ca^{2+} pump, which is insensitive to BHQ, and is bound by the extracellular EGTA.

4. Reconstitution with coelenterazine n. Depletion solution is removed and 0.2 ml of reconstitution solution is added to the well. Then, 1 µl of coelenterazine n 200 µM is added and gently mixed. The plate should then be left in the dark at room temperature for at least 1 hour. Reconstitution of aequorin with coelenterazine n is much slower than with wild type coelenterazine (the half-time in vitro is 5 h compared to 22 min for the wild-type, Shimomura et al., 1993). This means that longer times of reconstitution will give an almost linear increase in the total luminescence output. A very important factor is also to keep low the temperature of reconstitution. Increasing the temperature to 37 °C during reconstitution decreases by near one order of magnitude the total

luminescence of the sample, probably because of increased aequorin consumption.

5. Luminescence experiment. After reconstitution, the glass coverslip is transferred to the cell chamber of the luminometer and light recording is started. Extracellular medium containing EGTA 0.5 mM is perfused for 2-5 min to wash BHQ and release the inhibition of SERCA. Then, usually, extracellular medium containing 1 mM Ca^{2+} is perfused to refill the ER in order to perform afterwards a particular experiment (see below).

6. Cell lysis. In order to calibrate the data obtained in terms of $[Ca^{2+}]_{ER}$, we need to know the total amount of luminescence that can be emitted by the sample (see below). For that, at the end of every experiment it is essential to perfuse lysis solution containing detergent and excess Ca^{2+} to measure all the remaining aequorin luminescence.

Calibration of luminescence data

The calibration curve of mutated aequorin reconstituted with coelenterazine n has been determined at both 22 °C and 37 °C (Montero et al., 1997a; Barrero et al., 1997). Fig. 3 shows both calibration curves in a typical drawing of log L/Lmax against pCa (-log $[Ca^{2+}]$). L is the luminescence (counts per second) of a sample of mutated aequorin reconstituted with coelenterazine n and placed in a medium with a given $[Ca^{2+}]$, and Lmax is the total luminescence that can be recorded from the sample. L/Lmax is therefore the fraction of the total luminescence recorded from the sample in 1 s. For every $[Ca^{2+}]$, a particular fraction of the total aequorin luminescence is emitted per second, and the calibration curves show that this fraction increases rapidly with the $[Ca^{2+}]$. It is easy to understand now that the rate of aequorin consumption at high $[Ca^{2+}]$ is so fast that no stable measurement is possible. In fact, when log L/Lmax approaches the value -1, 10% of aequorin is consumed every second. We should remark also that crude luminescence values have no meaning in terms of $[Ca^{2+}]$ until they are transformed into ratios L/Lmax. The final calibrated results are therefore independent of the total amount of recon-

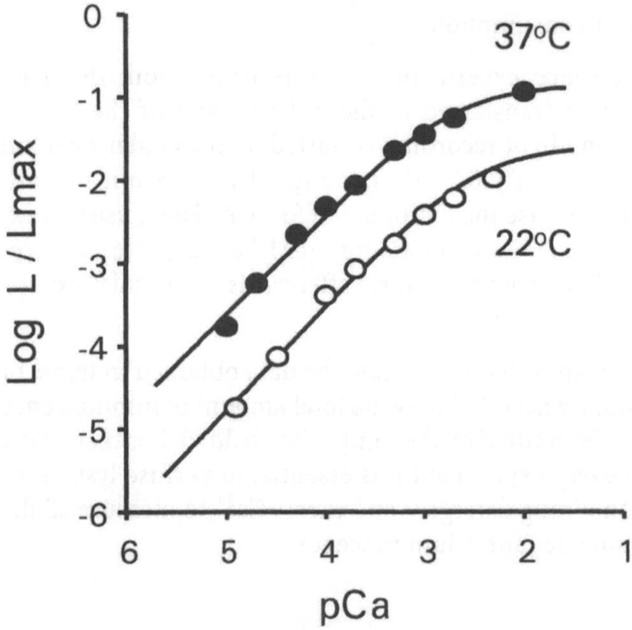

Fig. 3. Calibration curves of mutated aequorin reconstituted with coelenterazine n at both 22 °C and 37 °C. The curves were calculated as the best fittings of the algorithm (see text) to the experimental data. See Montero et al., 1997, Barrero et al., 1997, for more details on the calibration.

stituted aequorin in the sample. On the other hand, as aequorin is being progressively consumed, the value of Lmax (the total remaining luminescence of the sample) decreases continuously during the experiments. This leads to apparent paradoxes when compared with Ca^{2+}-sensitive fluorescent dyes. For example, a steady-state $[Ca^{2+}]_{ER}$ is not reached when the luminescence is at steady-state, but when luminescence decreases at a rate such that the ratio L/Lmax stays constant (see Fig. 4).

Figure 4 shows a detailed analysis of two experiments performed at 22 °C (left) and 37 °C (right). HeLa cells expressing ER-targeted mutated aequorin were reconstituted with coelenterazine n as detailed above, and 1 mM Ca^{2+} was added as indicated. The lower panels show the records of luminescence (counts per second) obtained in every experiment. Addition of Ca^{2+} triggers a rapid increase in luminescence, which reaches a peak and then decreases rapidly down to near background

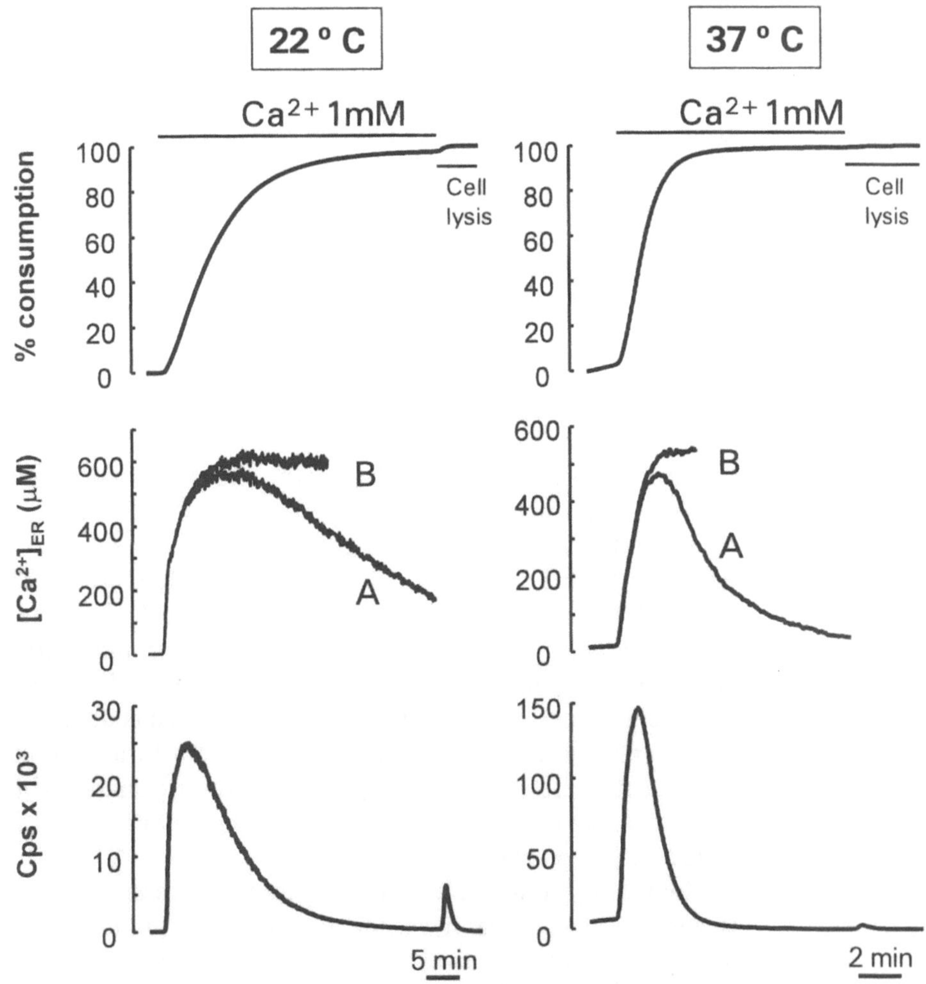

Fig. 4. Analysis of aequorin luminescence experiments performed at 22 °C and 37 °C. Lower panels, crude luminescence records. Middle panels, A: calibrated $[Ca^{2+}]$; B: calibrated $[Ca^{2+}]$ discarding the last 2% of aequorin luminescence. Upper panels, percentage of aequorin consumption during the experiments.

levels. Most of aequorin is consumed in this peak, because final cell lysis induces a quite small luminescence peak, which contains only about 1-2% of the total luminescence. Crude records of luminescence are stored in computer files that contain data sampled every 50 ms. To transform these data in $[Ca^{2+}]$, a computer program averages them 20-fold to obtain a data rate of 1 per second, subtracts the background and calculates the fractions L/Lmax at every point during the experiment. L is the luminescence value at every point (minus the background) and Lmax is the integral of luminescence (minus the background) from that point to the end of the experiment. L/Lmax values are then transformed into $[Ca^{2+}]$ values using the following mathematical algorithm (Brini et al., 1995).

$$[Ca^{2+}](\text{in M}) = \frac{\text{ratio} + (\text{ratio} \cdot K_{TR}) - 1}{K_R - (\text{ratio} \cdot K_R)},$$

$$\text{where ratio} = \left(\frac{L}{\text{Lmax} \cdot \lambda}\right)^{1/n}$$

The values for the parameters of the algorithm that should be used to calculate $[Ca^{2+}]$ in experiments using ERmutAEQ reconstituted with coelenterazine n are:
At 22 °C, $K_R = 5 \cdot 10^7$; $K_{TR} = 159.700$; $n = 1.271$; $\lambda = 0.02512$
At 37 °C, $K_R = 8.47 \cdot 10^7$; $K_{TR} = 165.600$; $n = 1.2038$; $\lambda = 0.138$

This algorithm was derived from a mathematical model proposed originally to explain from a molecular point of view the Ca^{2+}-dependence of aequorin luminescence (Allen et al., 1971), but it can be used as a simple mathematical transformation independent of the model. In fact, any other transformation able to fit correctly the experimental data in the calibration (Fig. 3) would be just as good.

Regarding the significance of the parameters in the model, it is worth indicating that n was the number of Ca^{2+}-binding sites in the model, and has a value of near 3 in native aequorin. The smaller value of n observed here is mainly due to the mutation of one of the Ca^{2+}-binding sites, and corresponds with a decrease in the slope of the calibration curve. The parameter λ is the rate constant for aequorin consumption at saturating $[Ca^{2+}]$. This parameter was not included in the original description of the algorithm (Brini et al., 1995), because the maximum rate constant of native aequorin reconstituted with wild-type coelenterazine is 1.0 s^{-1}. Reconstitution with coelenterazine n reduces considerably the maximum rate constant, and this allows the recording of high $[Ca^{2+}]_{ER}$ values with little aequorin consumption, particularly at 22 °C.

Fig. 4, middle panels, curve A, show the $[Ca^{2+}]$ values obtained by applying directly the algorithm to the experimental data. Fig.4, upper panels, show the normalized integral of aequorin luminescence during refilling of the ER, which indicates the percentage of aequorin consumed at every point during the experiment and reflects also the rate of consumption. We can see in the middle panels that refilling of the ER reaches $[Ca^{2+}]_{ER}$ values around 500-600 μM at every temperature. Aequorin consumption is quite fast at these $[Ca^{2+}]$ and the percentage of consumption increases rapidly, particularly at 37 °C. Comparing middle and upper panels of Fig. 4, we can observe that when the percentage of consumption gets above 80%, $[Ca^{2+}]_{ER}$ apparently starts to decrease. This phenomenon is due to the presence of a small amount of aequorin in a compartment with lower $[Ca^{2+}]$ (Montero et al., 1997b). When most of aequorin in high $[Ca^{2+}]$ compartments has been consumed, this compartment gains relative importance and tends to dominate the calibrated $[Ca^{2+}]_{ER}$ signal. The percentage of this compartment can be estimated from the amount of aequorin luminescence released by lysing the cells at the end of the experiment, when luminescence has returned to near-background levels. In the experiments shown, this was around 1-2%. In fact, if this aequorin placed in a low-Ca^{2+} environment is not taken into account in the calculations, the artifactual decrease in $[Ca^{2+}]_{ER}$ is no longer observed. Fig. 4, middle panels, curve B, show that $[Ca^{2+}]_{ER}$ reaches a stable steady-state when a value of Lmax of around 98% of that obtained experimentally is used for the calculations. This is a good example of the ability of aequorin to detect heterogeneities in $[Ca^{2+}]$. In this case, aequorin consumption in high $[Ca^{2+}]$ areas allows the detection of a small compartment (2% of total) with much lower $[Ca^{2+}]$. On the other hand, these kinds of experiments also constitute evidence that most of the ER (98% in HeLa cells) has a nearly homogeneous free $[Ca^{2+}]$ around 500 μM. The presence of compartments with grossly different free $[Ca^{2+}]$ would have been easily detectable by the presence of different rates of consumption along the refilling process.

Comparing the data obtained at both temperatures, the main technical difference is that the maximum rate of aequorin consumption (λ) increases about 5-fold at 37 °C (see also Fig. 3). This reduces the time that $[Ca^{2+}]_{ER}$ can be recorded at steady-state at 37 °C, even though the steady-state $[Ca^{2+}]_{ER}$ level is the same at both temperatures. While at 22 °C steady-state $[Ca^{2+}]_{ER}$ values can be monitored for 20-30 min, at 37 °C the measuring time at the same $[Ca^{2+}]_{ER}$ values is reduced to 3-5 min. Increasing this time would require introduction of a new mutation in the aequorin molecule to reduce further its Ca^{2+} affinity.

We should finally dedicate a few words here to the problem of the steady-state value of $[Ca^{2+}]_{ER}$. Using ERmutAEQ and coelenterazine n, we have obtained values for resting $[Ca^{2+}]_{ER}$ that range from 300 µM in cerebellar granule neurons to 500-600 µM in HeLa cells and 800 µM in pituitary GH_3 cells (Alonso et al., 1998). Previous work using native coelenterazine and either wild-type or mutated aequorin had reported $[Ca^{2+}]$ values that were grossly underestimated due to fast aequorin consumption and to heterogeneity of $[Ca^{2+}]$ within the ER (Montero et al., 1995, 1997b; Meldolesi and Pozzan, 1998). On the other hand, fluorescent dyes usually report $[Ca^{2+}]_{ER}$ values around 100-300 µM (Meldolesi and Pozzan, 1998; Mogami et al., 1998). Although somewhat lower, these values are in the same range as those obtained with ER-targeted aequorin. Moreover, difficulties in calibration and the presence of some dye in other organelles with low $[Ca^{2+}]$ such as mitochondria may lead to some underestimation of the $[Ca^{2+}]_{ER}$ values obtained with fluorescent dyes. On the contrary, targeting of aequorin is extremely precise, aequorin is hardly sensitive to Mg^{2+} (acts only as a competitive inhibitor, but does not trigger luminescence) and pH, and only a gross heterogeneity in $[Ca^{2+}]_{ER}$ among different regions of the ER would be expected to affect the measurement. As we show above, the ER-compartment containing low $[Ca^{2+}]$ in HeLa cells amounts only 1-2% of the total aequorin space, and correction for this space does not modify significantly the steady-state $[Ca^{2+}]_{ER}$ values.

Results

As mentioned above, the technique can be used with many cell types, even primary cultures, provided that the preparation is highly pure and that it is possible to express the ERmutAEQ gene. The purity of the preparation is very important here because it is not possible to separate the signal coming from dif-

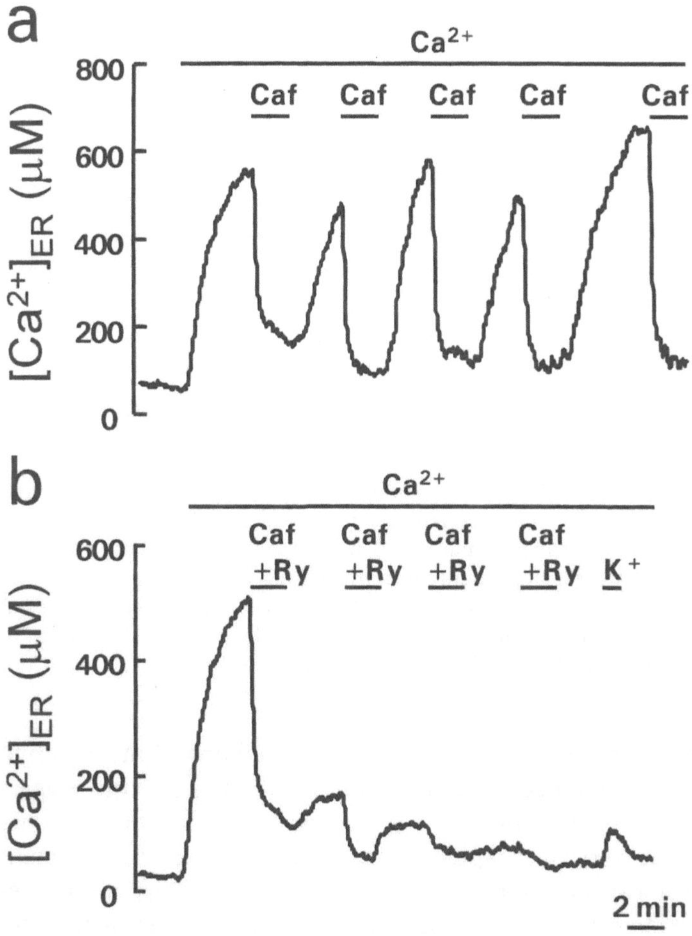

Fig. 5. Use dependent inhibition of ER refilling by ryanodine in chromaffin cells. The ER was refilled by perfusing with medium containing 1 mM Ca^{2+}. Then, several consecutive stimulations with 50 mM caffeine were performed as indicated, either in the absence (**a**) or in the presence (**b**) of 10 μM ryanodine. In panel **b**, standard medium containing 70 mM KCl (replacing an equimolar amount of NaCl) was perfused when indicated (K^+). Reprinted from The Journal of Cell Biology (1999), 144: 241-254 by Copyright permission of The Rockefeller University Press.

ferent cell types, and the intensity of the expression may also be different in different cells. We have measured $[Ca^{2+}]_{ER}$ in some preparations that can be obtained highly pure, such as chromaffin cells and cerebellar granule cells. Fig. 5a shows a typical experiment performed in chromaffin cells (Alonso et al., 1999).

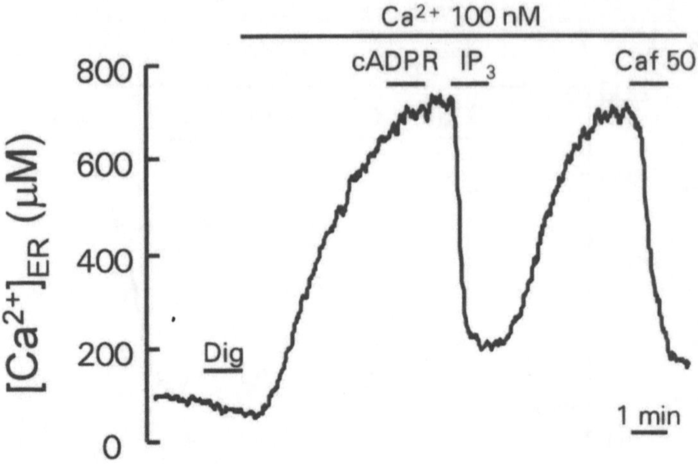

Fig. 6. Effects of cADPR, $InsP_3$ and caffeine on $[Ca^{2+}]_{ER}$ in permeabilized chromaffin cells. Cells were permeabilized by perfusion with 20 μM digitonin in intracellular medium for 1 min as indicated. Then the ER was refilled by perfusion with intracellular medium containing 100 nM Ca^{2+} (buffered with EGTA). Finally, either 5 μM cADPR, 2 μM $InsP_3$ or 50 mM caffeine were perfused as indicated. Reprinted from The Journal of Cell Biology (1999), 144: 241-254 by Copyright permission of The Rockefeller University Press.

Addition of extracellular Ca^{2+} leads to refilling of the ER up to values around 600 μM. Subsequent addition of caffeine triggers a rapid decrease in $[Ca^{2+}]_{ER}$, and washing of caffeine allows $[Ca^{2+}]_{ER}$ to recover rapidly to the steady-state values. This manoeuver can be repeated many times, and $[Ca^{2+}]_{ER}$ always recovers to steady-state values. This repetitive recovering is extremely important as a test that aequorin consumption does not affect significantly the $[Ca^{2+}]_{ER}$ values obtained in these cells. Panel b of Fig 5 shows a similar experiment but performed in the presence of ryanodine, an inhibitor of ryanodine receptors that locks open the ryanodine-sensitive Ca^{2+}-channels in the ER in a use-dependent manner when activated by caffeine. We can see that the first addition of caffeine triggers the same $[Ca^{2+}]_{ER}$ decrease, but then the ER is progressively unable to refill during the rest of the experiment. The physiological interpretation is that the ryanodine-sensitive Ca^{2+} channels remain open after binding of ryanodine and this makes ER refilling impossible. However, it is necessary to show first that the lack of recovery of $[Ca^{2+}]_{ER}$ is not an artifact due to aequorin consumption. For that, we compare this experiment with the control above, and measure the percent of aequorin consumption at the end of the experiment in both cases. In these experiments, aequorin consumption at the end of each experiment was 81.5% in panel A and 48% in panel B. Therefore, we can now conclude confidently that the lack of refilling in the experiment with ryanodine (panel B) is not due to aequorin consumption.

The technique also allows us to perform experiments with permeabilized cells. The protocol here is identical to that described above until the glass coverslip is placed in the luminometer and BHQ is washed. At that point, instead of perfusing extracellular medium containing 1 mM Ca^{2+}, we perfuse intracellular medium containing 0.5 mM EGTA and 20 μM digitonin for 1 min. Other cells may require use of higher digitonin concentrations (e.g. 100 μM for HeLa cells). Then, intracellular medium containing 100nM Ca^{2+} (buffered with EGTA) is perfused to refill the ER. Fig. 6 shows that the ER is refilled at a similar rate to that in intact cells (Fig.5) and reaches similar steady-state $[Ca^{2+}]_{ER}$ levels. At that point, we can test the effect of intracellular Ca^{2+} mobilizers such as cyclic ADP ribose (cADPR) or inositol 1,4,5-trisphosphate ($InsP_3$). These experiments on permeabilized cells may be particularly useful when we want to measure precisely the response of the ER to a certain intracellular agent.

Troubleshooting

Problems may appear in some steps of the protocol and most of them converge in a final poor luminescence yield. If cell lysis triggers little or no luminescence, we may have one of the following problems:

- Little or no expression of ERmutAEQ. An alternative way to test the expression of the protein is to perform immunofluorescence against the HA1 epitope. The anti-HA1 monoclonal antibody 12CA5 (Boehringer, see Montero et al., 1995 for details) is most adequate for this. A diffuse non-nuclear reticular pattern is typical of ER localization (Montero et al., 1995). We should say, however, that light detection from reconstituted aequorin is usually more sensitive than immunofluorescence, which is instead essential to prove the localization in the ER. If a problem of expression is suspected, we would recommend first trying expression in the desired cell culture of an easily detectable gene such as lac.

- If there is a good expression, but still poor luminescence yield, there may be a problem with cell attachment to the coverslip. Some cell types attach well to glass in a Ca^{2+}-containing medium but rapidly detach after the addition of EGTA. These cells should be attached to glass coverslips treated with special coatings (e.g. poly-lysine or similar) designed to facilitate attachment. The reduction of the EGTA concentration from 3 mM in the depletion solution to 0.5 mM in the reconstitution solution was designed to minimize this problem. Further reduction of the EGTA concentration in the reconstitution solution is still possible, and reasonable measurements can be performed even after reconstitution in nominally Ca^{2+}-free medium (with no EGTA added).
- Finally, the problem may be at the depletion and reconstitution step. Depletion may be incomplete for some reason, leading to aequorin consumption during reconstitution. Use of old BHQ or coelenterazine n solutions, or reconstitution at high temperature may both contribute to decrease of the luminescence yield.

Acknowledgements. Financial support from Dirección General de Enseñanza Superior (DGES, PM98-0142 and PB97/0474) and Junta de Castilla y León (VA 19/99) is gratefully acknowledged.

References

Allen DG, Blinks JR, Prendergast FG (1971) Aequorin luminescence: Relation of light emission to calcium concentration - A calcium-independent component. Science 195:996-998

Alonso MT, Barrero MJ, Carnicero E, Montero, M, Garcia-Sancho J, Alvarez J (1998). Functional measurements of [Ca^{2+}] in the endoplasmic reticulum using a herpes virus to deliver targeted aequorin. Cell Calcium 24:87-96.

Alonso MT, Barrero MJ, Michelena P, Carnicero E, Cuchillo I, García AG, García-Sancho J, Montero M, Alvarez J (1999) Ca^{2+}-induced Ca^{2+}release in chromaffin cells seen from inside the ER with targeted aequorin. J Cell Biol 144: 241-254

Barrero MJ, Montero M, Alvarez J (1997). Dynamics of [Ca^{2+}] in the endoplasmic reticulum and cytoplasm of intact HeLa cells. A comparative study. J Biol Chem 272:27694-27699.

Brini M, Marsault R, Bastianutto C, Alvarez J, Pozzan T, Rizzuto R (1995). Transfected aequorin in the measurement of cytosolic Ca^{2+} concentration ($[Ca^{2+}]_c$). A critical evaluation. J Biol Chem 270:9896-9903.

Meldolesi J, Pozzan T (1998). The endoplasmic reticulum Ca^{2+} store: a view from the lumen. Trends Biochem Sci 23:10-14.

Mogami H, Tepikin AV, Petersen OH (1998). Termination of cytosolic Ca^{2+} signals: Ca^{2+} reuptake into intracellular stores is regulated by the free Ca^{2+} concentration in the store lumen. EMBO J 17: 435-44

Montero M, Brini M, Marsault R, Alvarez J, Sitia R, Pozzan T, Rizzuto R (1995). Monitoring dynamic changes in free Ca^{2+} concentration in the endoplasmic reticulum of intact cells. EMBO J 14:5467-5475.

Montero M, Barrero MJ, Alvarez J (1997a). [Ca^{2+}] microdomains control agonist-induced Ca^{2+} release in intact HeLa cells. FASEB J 11:881-885.

Montero M, Alvarez J, Scheenen WJ, Rizzuto R, Meldolesi J, Pozzan T (1997b). Ca^{2+} homeostasis in the endoplasmic reticulum: coexistence of high and low [Ca^{2+}] subcompartments in intact HeLa cells. J Cell Biol 139:601-611.

Shimomura O, Kishi Y, Inouye S (1993). The relative rate of aequorin regeneration from apoaequorin and coelenterazine analogues. Biochem J 296:549-551

Suppliers

Thorn-Emi Electron Tubes Limited
Bury Street , Ruislip
Middlesex HA4 7TA
Phone +44 1895 630771
Fax +44 1895 635953

Molecular Probes Europe BV
Poort Gebouw, Rijnsburgerweg 10
2333 AA Leiden, The Netherlands
Phone: +31 71 5236850
Fax: +31 71 5233419
E-mail: eurorder@probes.nl

Ca^{2+} Measurement with Fluorescent Probes in the Endoplasmic Reticulum

ALDEBARAN M. HOFER

Introduction

The release of Ca^{2+} from intracellular storage compartments constitutes an essential component of most Ca^{2+} signaling cascades (Clapham, 1995). Thanks to the development and widespread use of sensitive probes for cytoplasmic Ca^{2+} such as fura-2 (Grynkiewicz 1985; Tsien, 1989), the calcium ion is likely the most intensively studied of the known second messenger systems. Cytoplasmic signals, however, are influenced by many factors (Pozzan et al., 1994). These include Ca^{2+} leaks and pumps at both the plasma membrane and in the endoplasmic reticulum (ER), and interactions with endogenous cellular buffers and transport by other organelles (i.e. mitochondria). Therefore, the examination of certain key features of the Ca^{2+} release and reloading process are best approached by monitoring the cation directly within Ca^{2+} stores.

Currently available methods for measuring free intraluminal ER [Ca^{2+}] fall into three categories:

1. targeted recombinant luminescent photoproteins (aequorin);
2. targeted fluorescent indicators based on green fluorescent proteins and calmodulin ('cameleons'; Miyawaki et al. 1997; Persechini et al. 1997), and
3. trapped low-affinity fluorescent dyes loaded as AM-esters.

Aldebaran M. Hofer, Harvard Medical School, Dept. Surgery, Brigham and Women's Hospital, West Roxbury VAMC, Surgery 112,1400 VFW Parkway, West Roxbury, MA, 02132, USA (*phone* +01-617-323-7700, ext. 5902; *fax* +01-617-363-5592; *e-mail* ahofer@rics.bwh.harvard.edu)

The first two approaches are described in chapters **XX** and **XX** of this volume, and elsewhere. Here the basic principles of the third approach, the use of compartmentalized low-affinity probes, will be discussed.

This method takes advantage of the tendency of AM-esters to accumulate in not only the cytoplasm, but also in subcellular compartments such as the ER. By using dyes with low-affinity for Ca^{2+} it is possible to monitor changes in free [Ca^{2+}] directly in cellular Ca^{2+} storage pools (Hofer and Machen, 1993).

The procedures involved (represented schematically in Figure 1) are relatively simple. Cells are incubated with the AM-ester of the appropriate fluorescent probe, which generally results in dye uptake into both organelles and the cytoplasm. The next step is to eliminate the cytoplasmic indicator, which other-

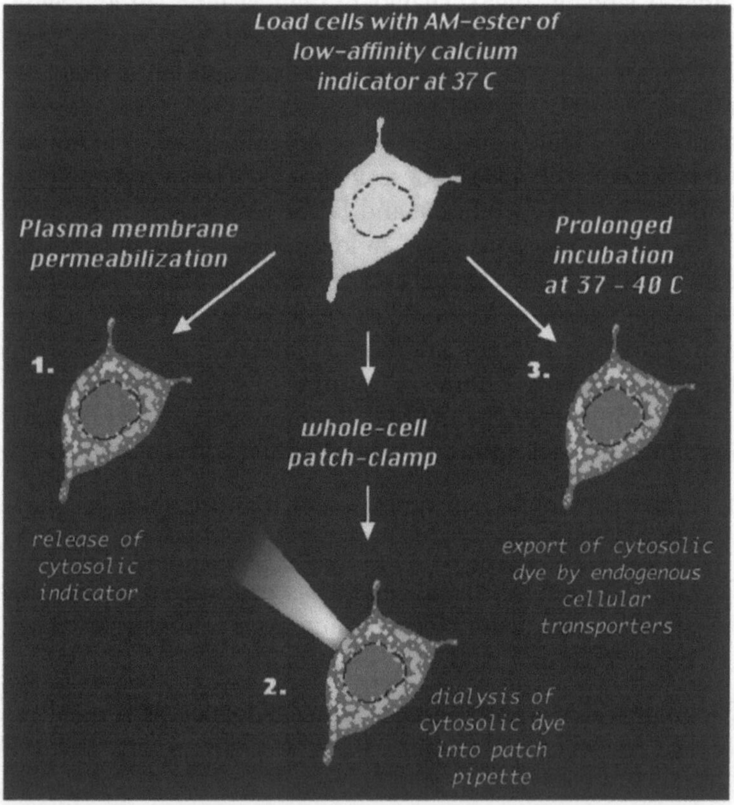

Fig. 1. Schematic representation of three different basic approaches for using compartmentalized dyes to measure intraluminal ER [Ca^{2+}].

wise produces high background levels of fluorescence that obscure the signals of interest. This can be accomplished either: 1) by permeabilizing the plasma membrane, 2) by dialysis with a patch pipette, or 3) for some cells, under particular conditions, by simply prolonging the incubation period. Fluorescence measurements of the residual dye are then performed as for the cytoplasm.

It should be emphasized that AM-esters of Ca^{2+} indicators are known to accumulate in a variety of organelles (mitochondria, secretory granules; Gerasimenko et al, 1996), as well as subcompartments of the ER (the nuclear envelope; Petersen et al, 1998). While this represents a simple way of introducing Ca^{2+} probes into these organelles (as outlined in other chapters of this volume), it can also seriously complicate the measurement of $[Ca^{2+}]$ in a given subcellular compartment (Hofer and Machen, 1994; Hofer and Schulz, 1996). With careful consideration of this fact, however, an abundance of information on ER Ca^{2+} handling can be obtained from the unique perspective of the internal store lumen.

Materials

Selection of low-affinity Ca^{2+} indicators

Because $[Ca^{2+}]$ in the ER has been estimated to range from 60 μM to around 800 μM (Hofer and Schulz, 1996; Alonso et al. 1998) it is necessary to use low-affinity dyes in order to monitor dynamic $[Ca^{2+}]$ changes. A wide variety of dyes with different spectral properties have been synthesized. Most are available from Molecular Probes Inc. As seen from Table 1, both ratiometric (e.g. mag-fura-2, mag-indo-1, fura-2-ff) and single wavelength indicators (e.g. fluo-3-ff, mag-fura-red) are available. The type of indicator to be used is most often dictated by the measuring system (i.e. confocal microscope vs. ratio imaging systems). Ratiometric probes offer obvious advantages, but as is the case for their high-affinity counterparts (fura-2, indo-1 etc.), the existing probes are excited in the UV part of the spectrum, requiring UV compatible microscope optics. Figure 2 shows the excitation spectrum of the ratiometric indicator, mag-fura-2, as measured from organelle-trapped dye of intact cells on the microscope stage. Be aware that the spectra of dyes loaded intracellularly are often different when measured in situ on the microscope stage, compared to in a cuvette.

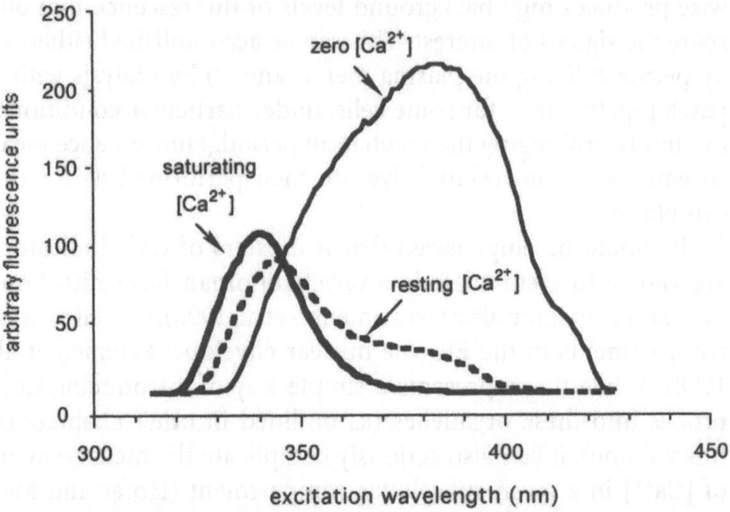

Fig. 2. Excitation spectrum (490 nm em.) of compartmentalized mag-fura-2 in an intact BHK-21 fibroblast measured on the microscope stage. The spectrum of organelle-trapped dye when stores were fully loaded with Ca^{2+} ("resting [Ca^{2+}]") is compared to the spectrum when stores had been emptied with ionomycin in Ca^{2+} -free solutions ("zero [Ca^{2+}]"), and when excess Ca^{2+} (10 mM) was added in the presence of ionomycin ("saturating [Ca^{2+}]"). Although mag-fura-2 normally has a large Ca^{2+}-dependent fluorescence change at 330 nm when measured in a cuvette (see Raju et al., 1989, and Molecular Probes Handbook), this change is largely attenuated on the microscope stage. This is due principally to the reduced light transmission by the optical elements of measuring system and because ionophores do not equilibrate internal and external [Ca^{2+}] to yield a saturating intraluminal [Ca^{2+}].

Some newer dyes excited by longer wavelengths (550–580 nm) have also recently become available (Molecular Probes Inc.). These include Rhod-5N and X-rhod-5N, with K_ds for Ca^{2+} around 320 μM, 350 μM respectively. These indicators have not yet been tested for the purpose of monitoring intraluminal ER Ca^{2+}, but their apparent lack of Mg^{2+} sensitivity and very low Ca^{2+} affinity may prove useful for some applications.

Table 1. Low-affinity fluorescent indicators that have been used to measure [Ca^{2+}] in organelles. Dyes are available from Molecular Probes Inc. unless otherwise noted.

Indicator	Kd for Ca2+ (μM)	excitation (nm) emission (nm)	comments
mag-fura-2 ('furaptra')	53	345/375 ex. 510 em.	ratiometric
mag-fura-5	28	340/380 ex. 510 em.	ratiometric
fura-2-ff	35	340/380 ex. 510 em	ratiometric; available from Teflabs
mag-indo-1	32	351 ex. 405/485 em.	ratiometric (emission)
mag-fura-red	17	488 ex. 630 em.	fluorescence decreases upon Ca2+ binding
fluo3-ff	41	515 ex. 530 em.	available from Teflabs; fluorescence increases upon Ca2+ binding
Oregon Green BAPTA-5N	20	492 ex. 521 em.	excited efficiently by 488 nm laser line; fluorescence increases upon Ca2+ binding

Stock solutions

Fluorescent indicators

Dissolve the AM-ester form of the dye in the appropriate volume of anhydrous DMSO to make a 5 - 10 mM stock solution. Divide this into small aliquots (10-20 μl each) and keep dry in the freezer (-20 °C) until use.

Other materials

Other reagents that may be needed, depending on the protocol to be employed include:

- EGTA stock solution (0.5 M, prepared accurately, dissolved in KOH and pH adjusted to 7.2)
- NTA (nitrilotriacetic acid) for in situ calibration [optional]: used at a final concentration of 1 mM in the calibration solution

- $CaCl_2$ stock solution: 1 M in distilled water
- Ca^{2+} ionophores: Ionomycin or 4Br-A23187 stock solutions: 10 mM in DMSO

Ionomycin is an electroneutral carrier of divalent cations with good selectivity for Ca^{2+} over Mg^{2+}. 4Br-A23187 is a non-fluorescent derivative of the Ca^{2+} / Mg^{2+} ionophore A23187. Keep in mind that both are effectively divalent cation / proton exchangers. Thus their efficiency in equilibrating Ca^{2+} in a given organelle also depends on the pH of that compartment.

Experiments in permeabilized cells

Permeabilizing agents

Excellent results have been obtained with streptolysin O (Ahnert-Hilger and Weller, 1998) a pore-forming toxin, which renders the plasma membrane permeable to relatively large molecules (up to M.W. = 150 kD). However, we have observed that there is much variability in the quality of commercially available preparations. Alpha toxin is another pore-forming toxin from the bacteria *Staphylococcus aureus* that has the advantage of making relatively small perforations in the plasma membrane (MW < around 3000). This is sufficiently large to let out indicator dyes from the cytoplasm (Steenberger and Fay, 1996) while allowing most intracellular constituents to be retained. A specific, unidentified plasma membrane 'receptor' allows this toxin to be effective at low concentrations in many cell types (Ahnert-Hilger and Weller, 1998). Digitonin is a reliable and inexpensive alternative that produces holes in the plasma membrane similar in size to those of streptolysin-O. Both of these permeabilizing agents rely on interactions with cholesterol (enriched in the plasma membrane) for their action. The protocol for the use of digitonin will be described below in detail.
Digitonin stock solution: 5 mg / ml in distilled water.

Solutions needed for the experiment

- Standard physiological buffer appropriate for the cell type being studied (e.g. Ringer's solution).
- KCl rinse solution (in mM): 125 KCl, 25 NaCl, 10 HEPES, 0.5 $MgCl_2$, pH 7.25.

Note: For experiments in permeabilized cells, a nominally Ca^{2+}-free KCl rinse solution and an intracellular-like buffer designed to approximate the cytosolic milieu will be necessary.

- Intracellular-like buffer (in mM): 125 KCl, 25 NaCl, 10 HEPES, 0.5 $MgCl_2$, 0.5 Na_2ATP, 200 μM $CaCl_2$, 500 μM EGTA, pH 7.25.
- Permeabilization solution: consists of the intracellular-like buffer supplemented with 5 μg/ml digitonin.

A few general comments about the composition of the intracellular-like buffer are in order: First, we have noticed that organelles tend to maintain their delicate reticular structure somewhat better when an impermeant osmolyte (e.g. sucrose or mannitol, 40-50 mM) is partially substituted for KCl in the intracellular-like solution (Hofer and Machen, unpublished observations). Similar preservation of organelle morphology was also observed when a portion of the Cl^- in the intracellular buffer was substituted with an impermeant anion such as gluconate (added for example as K gluconate, leaving 25 mM NaCl in the solution). These substitutions have no apparent effect on the kinetics of release or reloading of intracellular Ca^{2+} stores, however (Hofer et al., 1996).

Second, the combination of Ca^{2+} and EGTA in the above recipe will yield a calculated free [Ca^{2+}] of approx. 100 nM, which closely matches that of cytoplasmic [Ca^{2+}] at rest (Bers et al. 1994). Be aware, however, of the many factors that can affect the [Ca^{2+}] in EGTA buffers, such as [Mg^{2+}], [ATP], ionic strength, and most notably the pH of the solution. The latter has a very powerful effect on the resulting free [Ca^{2+}]. BAPTA is a pH-insensitive (but more costly) alternative to EGTA. Note also that EGTA and BAPTA bind many other multivalent cations (La^{3+}, Gd^{3+}, Mn^{2+}, Zn^{2+} etc). Care must be used when adding these metal ions to EGTA- or BAPTA-containing solutions because 1) the free concentration of the added cation is not what you think it is, and 2) they displace Ca^{2+} on EGTA or BAPTA binding sites and alter the free [Ca^{2+}] in the solution.

Note: It is highly recommended that investigators obtain a computer program that allows the calculation of free [Ca^{2+}] in a solution containing multiple ligands. A particularly good program is MaxChelator, which can now be conveniently accessed through the internet (http://www.stanford.edu/~cpatton/maxc.html). This site is an excellent resource and is updated regularly. Users can register so as to be informed of major revisions in the program.

A number of recent studies have highlighted the importance of the choice of Ca^{2+} buffer (BAPTA vs. EGTA) as a determinant of the profile of the Ca^{2+} response (Tanimura and Turner, 1996; Montero et al, 1997; Hajnóczky and Thomas, 1997). Because $InsP_3$-receptors are sensitive to their local Ca^{2+} environments at the cytosolic face, changing buffering conditions can alter the release dynamics in subtle ways. Hajnóczky and Thomas (1997) showed, for example, that rapid chelation of Ca^{2+} exiting from the store via the $InsP_3$ receptor by BAPTA prevents Ca^{2+} oscillations. These investigators were able to reconstitute internal store [Ca^{2+}] oscillations in permeabilized cells by using unbuffered (or lightly buffered) intracellular-like solutions. Since the contaminating [Ca^{2+}] in a nominally Ca^{2+}-free solution can range from 1-10 μM, these solutions require special preparation.

To achieve physiological [Ca^{2+}] in an intracellular-like solution without using Ca^{2+} buffers, it is necessary to run the solution over a chelating column (Morgan and Thomas, 1998). Previously we have also used Calcium Sponge S (from Molecular Probes Inc.), which are essentially BAPTA-conjugated polystyrene beads, for this purpose. Solutions can be passed over a column of the insoluble beads. Alternatively, the solutions can be stirred with the beads, which are then removed by centrifugation. It is possible to regenerate the beads several times. This is an effective way to remove contaminating Ca^{2+} (and other metal ions), but can be expensive if large volumes (more than about 100 ml) are needed.

Procedure

Loading cells with dye

Add the AM-ester directly to cells at a concentration of 2 - 5 μM, and swirl gently. Cultured cells can be put back into the tissue culture incubator. Pluronic F-127, a dispersing agent, can be used to improve the solubility of the AM-ester, but does not appear to alter the specificity of loading.

Loading time

A variety of loading times have been reported by different investigators, ranging from 20 minutes to several hours. Optimal incubation times must be determined empirically for each cell type. For example, for experiments in permeabilized BHK-21 fibroblasts, best results were obtained by loading 20 min. at 37 °C with 2 μM of mag-fura-2-AM. Missiaen et al. (1995) reported that excessive dye loading (3.5 hours with 40 μM mag-fura-2-AM) altered the kinetics of $InsP_3$ induced release as measured by $^{45}Ca^{2+}$ fluxes in A7r5 smooth muscle cell line. Whether this is a consequence of increased luminal Ca^{2+} buffering or because of the toxic effects resulting from the hydrolysis of the AM-ester (Negulescu and Machen, 1990) is not known. Obviously, as is the case for all fluorescent indicators introduced intracellularly as AM-esters, overloading should be avoided.

Permeabilizing the plasma membrane with digitonin

In order to establish optimal conditions, one can initially follow the permeabilization process on the microscope stage. Organelles are also susceptible to the effects of permeabilizing agents, so the time of exposure and digitonin concentration should be minimized as much as possible. Temperature is another important variable, with much more rapid disruption of membranes occurring at 37 °C.

1. After dye loading, rinse cells briefly with KCl rinse solution.
2. Exchange KCl rinse solution with the permeabilization solution containing digitonin. After several minutes cells should begin to lose the dye from the cytoplasm, resulting in a large, rapid drop in fluorescence intensity (for the case of mag-fura-2, this can be monitored at the isoexcitation wavelength of 345 nm). The nucleus will be nearly free of fluorescence, while dye trapped in the periphery should have a "reticular" or punctate appearance (similar to that shown for the intact cell in Figure 3).
3. Switch immediately to intracellular buffer without digitonin as soon as most of the cells have become permeabilized (usually less than 5 minutes).

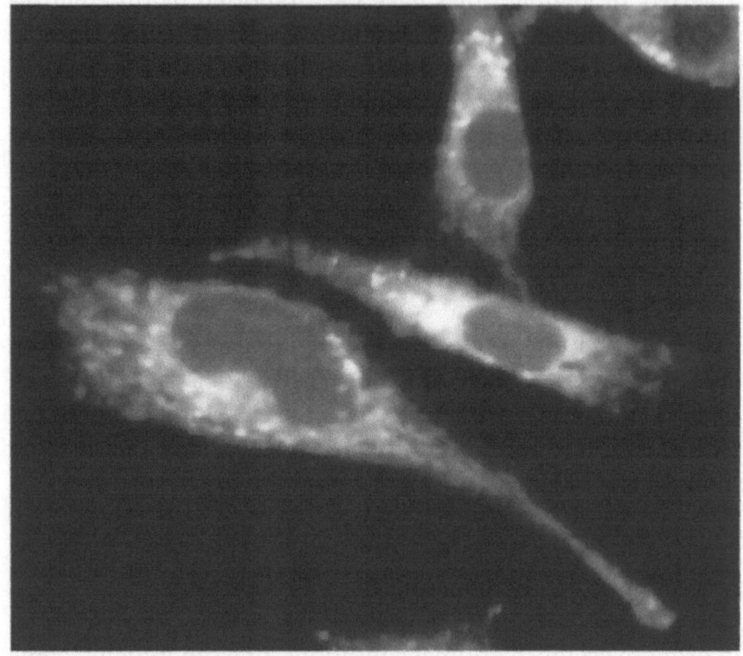

Fig. 3. Intact BHK-21 fibroblasts loaded with 2μM mag-fura-2-AM at 37 °C for 50 minutes. Fluorescence measured at Ca^{2+}-insensitive isoexcitation of the dye (345 nm) shows distribution typical of organelles.

Measurement of internal store $[Ca^{2+}]$

If the permeabilization process has been successful, the fluorophore will report a relatively high resting $[Ca^{2+}]$. The measurement of fluorescence from organelles of individual cells is similar to conventional procedures used for measuring cytoplasmic $[Ca^{2+}]$. Readers are referred elsewhere (Scheenen et al, 1998; Morgan and Thomas, 1998) for details of these protocols (setting of appropriate wavelengths, capture speed, gain, background etc.).

In our experiments we superfuse cells continuously with intracellular-like buffer during the experiment in order to: 1) wash away dye that may have leaked from the cells, and 2) also to maintain [ATP] and $[Ca^{2+}]$ bathing the stores at constant levels.

Determining the number and nature of intracellular Ca^{2+} stores

Different cells types possess diverse types of Ca^{2+} storing compartments (Fasolato et al., 1991), which can be defined functionally according to their susceptibility to specific Ca^{2+} releasing agents (e.g. $InsP_3$, SERCA inhibitors, and Ca^{2+} ionophores). It is important to characterize the nature and number of Ca^{2+} stores in the cell type under study for proper interpretation of the results (Hofer and Machen, 1994). For example, in BHK-21 fibroblasts, compartmentalized mag-fura-2 measures $[Ca^{2+}]$ changes in just one or two functional types of Ca^{2+} pool (Hofer et al. 1995). In this cell type sequestered Ca^{2+} can be released completely by the SERCA inhibitor thapsigargin and by 85% using a supramaximal dose of $InsP_3$ (6 μM). This does not necessarily mean that other organelles (e.g. mitochondria) do not accumulate dye, only that they do not store a detectable amount of Ca^{2+} as measured by the low affinity indicator. The situation can be much more complex in other cell types, where discrete or partially overlapping stores sensitive to SERCA inhibitors, ionophores, and other Ca^{2+} releasing agents ($InsP_3$, cADP ribose, sphingosine, etc.) may be present. It is important to note that the presence of multiple, independent, dye-containing compartments introduces a particular non-linearity in the fluorescence measurement, as described in detail by Hofer and Machen (1994) and Hofer and Schulz (1996).

Experiments in patch clamped cells

A number of investigators have taken advantage of the whole-cell patch clamp technique to dialyze indicator dyes from the cytoplasm of cells, allowing measurements of residual organelle-trapped fluorophore. This strategy has been used to perform simultaneous measurements of I_{CRAC} (a Ca^{2+} current activated by the depletion of internal stores) concurrently with the store $[Ca^{2+}]$ in mag-fura-2-loaded RBL-1 cells (Hofer et al. 1998a). With this general approach it has also been possible to study the precise relationship between ER $[Ca^{2+}]$ changes and $[Ca^{2+}]$ signaling events in the cytoplasm. The latter have been assayed indirectly as the activation of Ca^{2+}-sensitive Cl^- or K^+ currents, (Tse et al., 1994; Mogami et al., 1998), or directly by introducing free acid forms of high affinity Ca^{2+} dyes such as Fluo-3 via the patch pipette (Chatton et al. 1995).

Measurements in intact cells

Recently it has become evident that certain cell types can accumulate indicators preferentially in organelles with respect to the cytoplasm under conditions of prolonged dye loading (Golovina and Blaustein, 1997; Landolfi et al. 1998; Hofer et al. 1998b). This has allowed noninvasive measurements of store [Ca^{2+}] in single intact cells. This phenomenon apparently results from the extrusion of cytoplasmic dye across the plasma membrane, and is enhanced by raising the loading temperature to 40 °C.

Figure 3 illustrates the distribution of fluorescence in intact BHK-21 cells incubated with 2 µM mag-fura-2-AM for 50 minutes at 37 °C. The probe reported a high resting [Ca^{2+}] (not shown), and the appearance of the cells suggests that most (if not all) of the mag-fura-2 was localized to organelles. In fact, very little residual dye (0-15%) was released from BHK-21 cells loaded in this way following rupture of the plasma membrane with digitonin. As was the case for their permeablized counterparts, [Ca^{2+}] changes reported by the trapped dye were shown to occur exclusively in an agonist- and thapsigargin-releasable store (Hofer et al., 1998b). Best results have been obtained with subconfluent cells of relatively low passage number. BHK-21 cells are not unique in their ability to retain dye specifically in organelles; other cell types (e.g. smooth muscle cells, astrocytes) also apparently share this property (Golovina and Blaustein, 1997).

Figure 4 illustrates a typical response of intact BHK-21 fibroblasts following three sequential applications of the Ca^{2+} mobilizing agonist bradykinin (100 nM). It is interesting to note the rapid desensitization that occurs upon bradykinin addition, which is evident not only as a reduction in the magnitude of successive responses, but also as a refilling of stores in the continued presence of the agonist.

Calibration

Prior to this decade, one of the longstanding questions in the field of cell biology had been the determination of the resting free [Ca^{2+}] within the lumen of the ER. In past years, different techniques for measuring intraluminal ER [Ca^{2+}] yielded dramatically diverging estimates. For example, a value of 1–2 mM was obtained using recombinant targeted aequorins with Sr^{2+} as a surrogate (Montero et al. 1995) vs. ≈ 10 µM using null point titration techniques (Dawson et al.1995) vs. ≈200 µM using compartmentalized mag-fura-2 (Hofer et al. 1995). Careful analysis of the mag-fura-2 technique indicated that it potentially underestimates luminal [Ca^{2+}] (Hofer and Schulz, 1996),

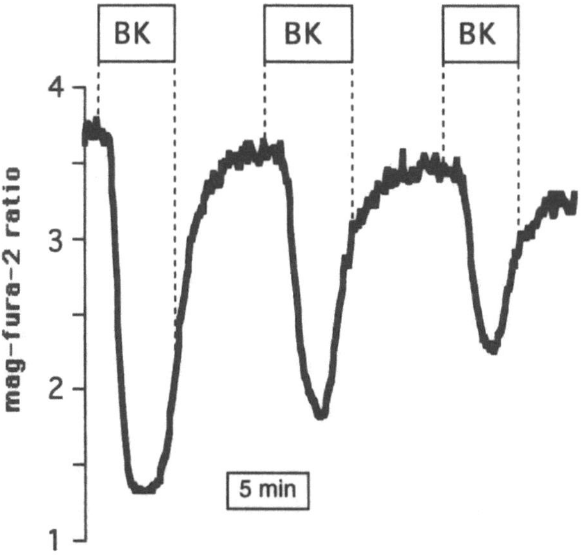

Fig. 4. 345 / 375 nm mag-fura-2 ratio measured in intact BHK-21 fibroblasts. Response to three sequential stimulations with 100 nM bradykinin. Note that the response rapidly desensitizes, allowing stores to refill in the continued presence of agonist.

while improvements in recombinant aequorin technology (such as the use of n-coelenterazine) showed [Ca^{2+}] to be somewhat lower (400–800 μM) than thought previously (Montero et al. 1997, Alonso et al., 1998). It seems that an informal consensus has finally been reached that places the average intraluminal ER [Ca^{2+}] at around 400 μM, and this appraisal was strengthened following independent reports of measurements using ER-targeted 'cameleons' (Miyawaki et al, 1997). These latter studies also indicated that there is considerable intercellular variation in the resting [Ca^{2+}], as was observed previously in mag-fura-2-loaded cells.

The conversion of fluorescence from organelle-trapped indicators into actual free [Ca^{2+}] is subject to all of the same difficulties that one may encounter during the calibration of high-affinity dyes. These have been described in detail previously and include susceptibility to heavy metals, pH, viscosity, ionic strength, bleaching and photodegradation (Roe et al., 1990; Scheenen et al. 1998). In addition, however, there are some special considerations for calibrating dyes in internal stores (see

Hofer and Schulz for detailed analysis). Specifically, it should be emphasized that the calibration cannot provide an absolute measure of $[Ca^{2+}]$ unless it is certain that dye is contained only in the organelle of interest. It is useful however, in that it can correct for the nonlinearity of the dye response.

There are three potential approaches for converting dye fluorescence into free $[Ca^{2+}]$:

- in vitro calibration
- R_{min} / R_{max} method for ratiometric dyes (or the corresponding technique for single wavelength indicators).
- in situ calibration with ionophores.

The in vitro calibration involves measuring a sample of the free acid form of the fluorophore in the presence of different Ca^{2+} concentrations on the stage of the microscope. These values are then compared to those obtained in living cells. As is the case with high affinity Ca^{2+} indicators, there are dramatic differences between the behavior of intracellular dye compared to probe in solution, thus this method provides only a first approximation of the $[Ca^{2+}]$ measured within organelles.

The second method is similar to that described for calibrating high affinity Ca^{2+} indicators, and involves the determination of the fluorescence of intraorganellar dye in situ in zero and saturating $[Ca^{2+}]$. Ionophores are used to equilibrate internal and external $[Ca^{2+}]$. For the case of a ratiometric indicator such as mag-fura-2 one can use the well-known "Grynkiewicz equation" first described by Tsien and collaborators:

$$[Ca^{2+}] = K_d * (S_{f2}/S_{b2}) * (R - R_{min})/(R_{max} - R)$$

where the K_d equals the dissociation constant of the probe, S_{f2} is the maximum fluorescence at 380nm (in zero Ca^{2+}), S_{b2} is the minimum fluorescence at 380nm (in saturating Ca^{2+}), R is the measured ratio, and R_{min} and R_{max} the minimum and maximum ratio in zero and saturating Ca^{2+}, respectively. Details of these procedures have been reported previously (Grynkiewicz, 1985).

A problem encountered with the above approach as applied to compartmentalized low-affinity dyes is the difficulty in obtaining a true value for R_{max}. The estimated K_d of ionomycin for Ca^{2+} is around 100 µM, while high Ca^{2+} concentrations

($\approx$10 mM) are required to saturate low-affinity probes such as mag-fura-2. A stratagem to determine R_{max} is described here that takes advantage of the fact that Mn^{2+} ions (also carried by ionophores) bind the dye with high affinity, and can be used as a surrogate for Ca^{2+} ions.

Figure 5 shows the **background subtracted** fluorescence intensity at 345 nm, 380 nm, and the corresponding 345 / 380 nm ratio from a permeabilized mag-fura-2-loaded cell. The cell was initially stimulated with $InsP_3$, and then treated with ionomycin in a Ca^{2+}-free / EGTA-containing solution to obtain R_{min} and S_{f2}. Conventionally, R_{max} and S_{b2} are next determined by treating the cell with a saturating [Ca^{2+}] (10 mM) in the presence of ionomycin. As illustrated in the figure, a relatively long period of time is required before the ratio approaches its maximum in 10 mM Ca^{2+}. It is evident that saturation has not been reached because subsequent addition of Mn^{2+} yields an additional decrease in fluorescence at 380 nm. Mn^{2+} quenches mag-fura-2 fluorescence at all wavelengths, but Mn^{2+} and a saturating [Ca^{2+}] cause the same degree of quenching at the Ca^{2+}-sensitive wavelength of 380nm. One can therefore approximate R_{max} by taking the value of the fluorescence intensity at 345nm immediately prior to Mn^{2+} addition (F_{345nm}), and dividing this by the Mn^{2+} -quenched intensity at 380nm (F_{380nm}) to yield R_{max}*. It is assumed that the 345 nm intensity will not have changed substantially during this time. Correspondingly, a more accurate value for S_{b2} (which is the same as F_{380nm}) can also be obtained. Using these substitutions (R_{max}* for R_{max}, and F_{380} for S_{b2}) in the preceding equation and applying a K_d of 53 µM, free [Ca^{2+}] within organelles of this particular cell was estimated to fall from a resting value of 489 µM to 36 µM following $InsP_3$ stimulation.

The third basic approach is the in situ determination, which relies on ionophore-mediated equilibration of luminal [Ca^{2+}] with a series of external calibration solutions of known [Ca^{2+}]. Suitable Ca^{2+} concentrations to be used in these solutions would be 0, 100, 200, 400, 600, and 800 µM Ca^{2+}. Ionomycin or 4Br-A23187 is added at a final concentration of 10µM. In case greater accuracy is desired, the free [Ca^{2+}] can be fixed precisely using NTA buffers according to the computer program described above. These procedures have been described in detail elsewhere (Hofer, 1998). An important advantage of this approach is that it is not necessary to know the K_d of the dye, which is influenced by many factors (viscosity, ionic strength etc.).

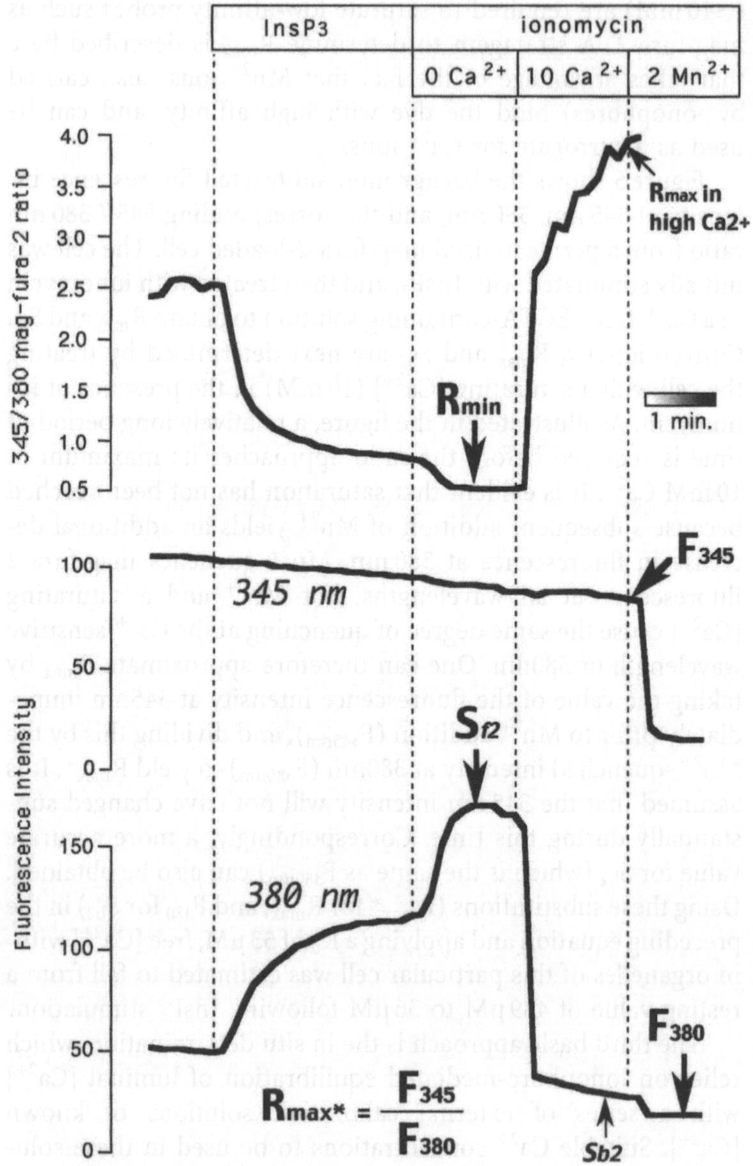

Fig. 5. Calibration procedure for ratiometric probe mag-fura-2 using Mn^{2+} as a "surrogate" for Ca^{2+}.

Pitfalls

Measurements using compartmentalized low-affinity indicators are subject to many of the same difficulties that plague cytoplasmic measurements of [Ca^{2+}]. The various factors that can influence the K_d of indicator dyes (alluded to above: protein concentration, viscosity, solvent polarity, etc.), have been discussed in detail elsewhere (Roe et al., 1990). A few other specific considerations follow:

- Many fluorescent Ca^{2+} indicators are sensitive (to varying extents) to changing pH. Appropriate controls are needed if a particular experimental manoeuver is suspected to alter intracellular or intraorganellar pH. For example, mitochondrial inhibitors can cause significant intracellular acidification that can potentially influence dye fluorescence. Be aware also that while certain membrane permeant weak acids or bases (e.g. flufenamic acid, acetate, CO_2, ammonium) may not influence the pH when added to a solution, they can nevertheless alter the pH within cellular compartments because only the acidic or basic species traverses membranes.

- Experimental agents added to the bath may themselves be fluorescent, or may screen the excitation or emission of the dye. In the case of ratiometric indicators, examination of the raw fluorescence data can often help to diagnose such artifacts. Physical or chemical interactions can also occur between the fluorophore and certain compounds, resulting in spurious fluorescence changes. Of special note in this regard are recent observations by Muschol et al. (1999) describing the influence of caffeine on the fluorescence of several Ca^{2+} indicators, including mag-fura-2.

- Some cell types possess large endogenous stores of zinc and other heavy metals. Virtually all fluorescent Ca^{2+} probes are susceptible to these contaminating ions, which frequently bind the dye molecule with far greater affinity than Ca^{2+} itself (Simons, 1993). This can result in quenching of dye fluorescence, reducing the sensitivity of the indicator towards Ca^{2+}. Alternatively, these ions can mimic Ca^{2+}, yielding a falsely high [Ca^{2+}].
  Low doses (10 μM) of the membrane-permeant heavy metal chelator, TPEN, can be used to assess the contribution of

heavy metals to the overall signal (Arslan et al, 1995; Snitsarev et al, 1996). For mag-fura-2, the ratio may or may not decrease following application of the chelator, while the individual fluorescence intensities will increase at both wavelengths as the TPEN competes away heavy metals bound to the dye. Note that higher concentrations of TPEN (50 μM or more) can chelate Ca^{2+} within internal stores, a fact which has been used to advantage as a means of controlling the free intraluminal ER [Ca^{2+}] independent of Ca^{2+} release (Hofer et al. 1998a).

- A frequent criticism aimed at the compartmentalized dye technique concerns the Mg^{2+} sensitivity of the more popular indicators such as mag-fura-2 and mag-indo-1. The affinity of these probes for Ca^{2+} is actually much higher than for Mg^{2+}. For the case of mag-fura-2, dramatic excursions in organelle [Mg^{2+}] (from 0 to 10 mM) are required to mimic the effect of a 100 μM change in intraluminal [Ca^{2+}] (Hofer and Schulz, 1996). Moreover, several studies have concluded that [Mg^{2+}] changes do not interfere significantly with the measurement of [Ca^{2+}] in $InsP_3$-sensitive Ca^{2+} storing compartments (Sugiyama and Goldman, 1995; Hofer and Machen, 1994; Hofer and Schulz, 1996). Nevertheless, it should not be assumed that [Mg^{2+}] changes will not be reported by the dye under some conditions. The recent introduction of probes (25) with reduced sensitivity to Mg^{2+} (for example fura-2-ff [London et al. 1994], fluo-3-ff) provide the possibility to control these potential artifacts.
- Compartmentalized dye within organelles of intact cells may gradually redistribute (for example throughout the elements of the secretory pathway) during long experiments. This problem is more pronounced when cells are maintained at 37 °C, and may result in a drift in the resting ratio over time.
- Considerable perturbations in subcellular architecture and organelle morphology can be induced by cell permeabilization (Hofer and Pozzan, unpublished observations).

Acknowledgements. The author is indebted to the many colleagues who have provided advice and encouragement throughout the years. Special thanks to Tullio Pozzan, Terry Machen, Irene Schulz, Silvana Curci, and the members of their respective laboratories for all their help.

References

Ahnert-Hilger G, Weller U (1998) Alpha-toxin and streptolysin-O as tools in cell biological research. In: J.E. Celis (editor)"Cell Biology: A Laboratory Handbook" Academic Press, New York, vol. 4: pp 103-110.

Alonso MT, Barrero MJ, Carnicero E, Montero M, Garcia-Sancho J, Alvarez J (1998) Functional measurements of $[Ca^{2+}]$ in the endoplasmic reticulum using a herpes virus to deliver targeted aequorin. Cell Calcium 24(2): 87-96.

Arslan P, Di Virgilio F, Betrame M, Tsien RY, and Pozzan T (1985) Cytosolic Ca^{2+} homeostasis in Ehrlich and Yoshida carcinomas. A new membrane-permeant chelator of heavy metals reveals that these ascites tumor cell lines have normal cytosolic free Ca^{2+}. J. Biol. Chem. 260: 2719-2727.

Bers DM, Patton CW (1994) A practical guide to the preparation of Ca^{2+} bufferes. Nuccitelli, R. (ed.) A practical guide to the study of calcium in living cells. In "Methods in Cell Biology" Vol. 40. Academic Press, San Diego.

Chatton J-Y, Liu H, and Stucki JW (1995) Simultaneous measurement of Ca^{2+} in the intracellular stores and cytosol of hepatocytes during hormone-induced Ca^{2+} oscillations. FEBS Letters 368: 165-168.

Clapham DE (1995) Calcium signaling. Cell, 80(2): 259-68.

Dawson AP, Rich, GT, Loomis-Husselbee, JW (1995) Estimation of the free $[Ca^{2+}]$ gradient across endoplasmic reticulum membranes by a null-point method. Biochem J. 310:371-374.

Fasolato C, Zottini M, Clementi E, Zacchetti D, Meldolesi J and Pozzan T (1991) Intracellular Ca^{2+} pools in PC-12 cells. Three intracellular pools are distinguished by their turnover and mechanisms of Ca^{2+} accumulation, storage, and release. J. Biol. Chem. 266: 20159-20167.

Gerasimenko OV, Gerasimenko JV, Belan PV, and Petersen OH (1996) Inositol trisphosphate and cyclic ADP-ribose-mediated release of Ca^{2+} from single isolated pancreatic zymogen granules. Cell 84(3): 473-480

Golovina, V.A. and Blaustein, M.P. (1997) Spatially and functionally distinct Ca^{2+} stores in sarcoplasmic and endoplasmic reticulum. Science 275(5306), 1643-8.

Grynkiewicz G, Poenie M, and Tsien RY (1985) A new generation of Ca^{2+} indicators with greatly improved fluorescence properties. J. Biol. Chem. 260, 3440-3450.

Hajnóczky G and Thomas AP (1997) Minimal requirements for calcium oscillations driven by the IP3 receptor. EMBO J. 16(12), 3533-3543.

Hofer AM, and Machen TE (1993)Technique for in situ measurement of calcium in intracellular $InsP_3$-sensitive stores using the fluorescent indicator mag-fura-2. Proc. Natl. Acad. Sci. USA. 90:2598-2602.

Hofer AM and Machen TE (1994) Direct measurement of free Ca^{2+} in organelles of gastric epithelial cells. Am. J. Physiol. 267: G442-G451.

Hofer AM, Schlue WR, Curci S, and Machen TE (1995) Spatial distribution and quantitation of free luminal $[Ca^{2+}]$ within the $InsP_3$-sensitive internal store of individual BHK-21 cells. Ion dependence of $InsP_3$-induced Ca^{2+} release and reloading. FASEB J. 9: 788-798.

Hofer AM and Schulz I (1996) Quantification of intraluminal free [Ca^{2+}] in the agonist-sensitive internal calcium store using compartmentalized fluorescent indicators: some considerations. Cell Calcium 20(3):235-242.

Hofer AM, Landolfi B, Debellis L, Pozzan T, Curci S (1998) Free [Ca^{2+}] dynamics measured in agonist-sensitive stores of single living intact cells: a new look at the refilling process. EMBO J. 17(7):1986-1996.

Hofer AM, Fasolato C, Pozzan T (1998) Capacitative Ca^{2+} entry is closely linked to the filling state of internal Ca^{2+} stores: a study using simultaneous measurements of ICRAC and intraluminal [Ca^{2+}]. J Cell Biol. 140(2):325-334.

Hofer AM (1998) Measurement of free [Ca^{2+}] changes in agonist-sensitive internal stores using compartmentalized fluorescent indicators. In "Methods in Molecular Biology: Calcium Signaling Protocols" (D.G. Lambert, editor) Humana Press, New Jersey, USA. pp249-265.

Landolfi B, Curci S, Debellis L, Pozzan T, Hofer AM (1998) Ca^{2+} homeostasis in the agonist-sensitive internal store: functional interactions between mitochondria and the ER measured in situ in intact cells. J Cell Biol. 142(5): 1235-43.

London RE, Rhee CK, Murphy E, Gabel S and Levy LA (1994) NMR-sensitive fluorinated and fluorescent intracellular calcium ion indicators with high dissociation constants. Am. J. Physiol. 266: C1313-C1322.

Miyawaki A, Llopis J, Heim R, McCaffery JM, Adams JA, Ikura M, Tsien RY (1997) Fluorescent indicators for Ca^{2+} based on green fluorescent proteins and calmodulin, Nature 388(6645): 882-7.

Missiaen L, Parys JB, DeSmedt H, Sienaert I, Henning RH, Casteels R (1995) Opening up Ca^{2+} stores with $InsP_3$. Nature 376:299-301.

Mogami H, Tepikin AV, Petersen OH (1998) Termination of cytosolic Ca^{2+} signals: Ca^{2+} reuptake into intracellular stores is regulated by the free Ca^{2+} concentration in the store lumen, EMBO J, 17(2):435-42.

Montero M, Brini M, Marsault R, Alvarez J, Sitia R, Pozzan T and Rizzuto R (1995) Monitoring dynamic changes in free Ca^{2+} concentration in the endoplasmic reticulum of intact cells. EMBO J. 14: 5467-5475.

Montero M, Barrero MJ, Alvarez J (1997) [Ca^{2+}] microdomains control agonist-induced Ca^{2+} release in intact HeLa cells. FASEB J. 11(11): 881-5.

Morgan AJ, Thomas AP (1998) Single cell and subcellular measurement of intracellular Ca^{2+} concentration ([Ca^{2+}]i). In "Methods in Molecular Biology: Calcium signaling protocols" (D.G. Lambert, editor) Humana Press, New Jersey, USA, pp. 93-123.

Muschol M, Dasgupta BR, Salzberg BM (1999) Direct interaction of caffeine with fluorescent calcium indicator dyes. In press. Biophys. J.

Negulescu PA and Machen TE (1990) Intracellular ion activities and membrane transport in parietal cells measured with fluorescent dyes. Academic Press San Diego.In Methods in Enzymology (S. Fleischer and B. Fleischer, eds) vol 192: pp. 38-81

Persechini A, Lynch JA, Romoser VA (1997) Novel fluorescent indicator proteins for monitoring free intracellular Ca^{2+} . Cell Calcium (3): 209-16.

Petersen OH, Gerasimenko OV, Gerasimenko JV, Mogami H, Tepikin AV (1998) The calcium store in the nuclear envelope. Cell Calcium 23(2-3):87-90

Pozzan T, Rizzuto R, Volpe P, Meldolesi J (1994) Molecular and cellular physiology of intracellular calcium stores. Physiol Rev 74(3):595-636.

Raju B, Murphy E, Levy LA, Hall RD, London RE (1989) A fluorescent indicator for measuring cytosolic free magnesium. Am. J. Physiol. 256:C540-C548.

Roe MW, Lemasters JJ, Herman B (1990) Assessment of fura-2 for measurements of free cytosolic calcium. Cell Calcium 11: 63-73

Romoser VA, Hinkle PM, Persechini (1997) A Detection in living cells of Ca^{2+}-dependent changes in the fluorescence emission of an indicator composed of two green fluorescent protein variants linked by a calmodulin-binding sequence. A new class of fluorescent indicators. J Biol Chem, 272(20): 13270-4

Scheenen WJJM, Hofer AM, Pozzan T (1998) Intracellular measurement of calcium using fluorescent probes. in "Cell Biology: A Laboratory Handbook" (J.E. Celis, editor) Academic Press, New York, pp. 363-374.

Simons T J B (1993) Measurement of free Zn^{2+} ion concentration with the fluorescent probe mag-fura-2 (furaptra). J. Biochemical and Biophysical Methods 27: 25-37.

Snitsarev VA, McNulty TJ, Taylor CW (1996) Endogenous heavy metal ions perturb fura-2 measurements of basal and hormone-evoked Ca^{2+} signals. Biophys. J. 71: 1048-1056.

Steenberger JM and Fay FS (1996) The quantal nature of calcium release to caffeine in single smooth muscle cells results from activation of the sarcoplasmic reticulum Ca^{2+} - ATPase. J. Biol. Chem. 271 (4):1821-1824.

Sugiyama T and Goldman WF (1995) Measurement of SR free Ca^{2+} and Mg in permeabilized smooth muscle cells with use of furaptra. Am. J. Physiol. 269:C698-C705.

Tanimura A, Turner RJ (1996) Inositol 1,4,5-trisphosphate-dependent oscillations of luminal [Ca^{2+}] in permeabilized HSY cells, J. Biol. Chem. 271: 30904-30908.

Tse FW, Tse A, Hille B (1994) Cyclic Ca^{2+} changes in intracellular stores of gonadotropes during gonadotropin-releasing hormone-stimulated Ca^{2+} oscillations. Proc. Natl. Acad. Sci. USA 91: 9750-9754.

Tsien RY (1989) Fluorescent indicators of ion concentrations. Methods in Cell Biology, 30: 127-156.

Ca^{2+} Measurements in Mitochondria

PAOLO PINTON, ROBERT DRUMMOND, PAULO MAGALHÃES, MARISA BRINI, ANNA CHIESA, TULLIO POZZAN, and ROSARIO RIZZUTO

Introduction

The contribution of mitochondria to intracellular Ca^{2+} signalling and the role of mitochondrial Ca^{2+} uptake, both in shaping the cytoplasmic response and controlling mitochondrial function, are currently areas of intense investigation. These studies rely to a large extent on the appropriate use of emerging techniques, coupled with judicious data interpretation. Calcium mitochondrial traffic takes place essentially through two pathways: i) an electrophoretic uniporter, allowing the accumulation of Ca^{2+} down its electrochemical gradient, and ii) an electroneutral antiporter that by exchanging Ca^{2+} with either Na^+ or H^+ prevents the attainment of an electrochemical equilibrium (Fig. 1).

Paolo Pinton, University of Padua, Department of Biomedical Sciences, Via G. Colombo 3, Padua, 35121, Italy
Robert Drummond, University of Massachusetts Medical Center, Biomedical Imaging Group, 373 Plantation St., Worcester, MA, 01605, USA
Paulo Magalhães, University of Padua, Department of Biomedical Sciences, Via G. Colombo 3, Padua, 35121, Italy
Marisa Brini, University of Padua, Department of Biochemistry, Via G. Colombo 3, Padua, 35121, Italy
Anna Chiesa, University of Ferrara, Department of Experimental and Diagnostic Medicine, Section of General Pathology, Via L. Borsari 46, Ferrara, 44100, Italy
Tullio Pozzan, University of Padua, Department of Biomedical Sciences, Via G. Colombo 3, Padua, 35121, Italy
✉ Rosario Rizzuto, University of Padua, Department of Biomedical Sciences, Via G. Colombo 3, Padua, 35121, Italy, University of Ferrara, Department of Experimental and Diagnostic Medicine, Section of General Pathology, Via L. Borsari 46, Ferrara, 44100, Italy (*phone* +39-0532-291361; *fax* +39-0532-247278; *e-mail* rzr@unife.it)

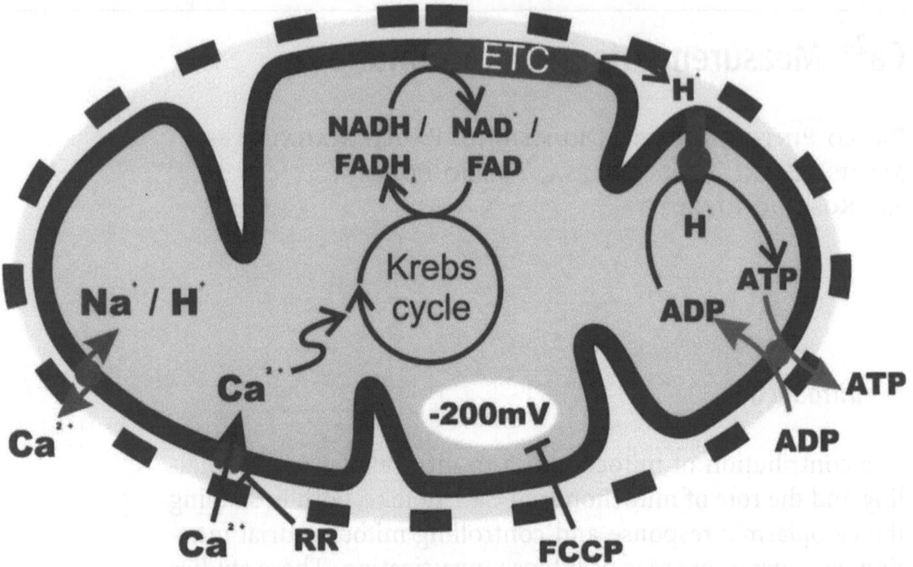

Fig. 1. Schematic representation of mitochondrial calcium homeostasis. Ca^{2+} entry takes place via a low affinity uniporter (which is inhibited by Ruthenium red), due to the high electronegative potential (which is collapsed by FCCP) in the mitochondrial matrix. Extrusion of Ca^{2+} takes place through an electro neutral antiporter (in exchange with either Na^+ or H^+). In the matrix, Ca^{2+} stimulates the activity of three Ca^{2+}-sensitive dehydrogenases of the Krebs cycle (NAD^+-isocitrate-, 2-oxoglutarate-, and pyruvate-dehydrogenase) thus promoting electron flow through the electron transport chain (ETC).

Unfortunately, although the biochemical properties of these transport pathways have been known for over three decades, there is still limited knowledge both at the molecular level and regarding the mechanisms that regulate the dynamics of this system. In this chapter we describe two methodological approaches to measure the concentration of Ca^{2+} within the mitochondrial context. In the first part we focus on the use of Rhod-2 (a fluorescent dye), while the second part is devoted to measurements using chimeric versions of aequorin (a naturally occurring protein).

Mitochondrial calcium measurements using Rhod-2

Over the past 15 years, fluorescent indicators for Ca^{2+} have become the dominant methodology for measuring the free cytosolic Ca^{2+} concentration ($[Ca^{2+}]_c$). There are a number of factors which make these indicators such powerful tools for measuring $[Ca^{2+}]$. First, they can be extremely sensitive to Ca^{2+}, having K_ds which range from ~100 nM to 50 μM (Haugland 1996). They also have excellent molecular specificity for Ca^{2+} over other ions. Second, they are capable of providing high temporal resolution pertaining to the cellular Ca^{2+} signal in a non-destructive manner. Furthermore, in combination with digital imaging microscopy, these fluorescent indicators can give entire images of ion concentration with resolution down to microns. Third, the development of membrane permeant acetoxymethyl (AM) esters of these fluorescent indicators, means that they are relatively easy to load into cells (Tsien et al. 1982). Additionally, they are equally applicable for studying Ca^{2+} in single cells as well as multi-cellular preparations (Tsien 1989). While the majority of studies using these indicators have focused primarily on measurement of cytosolic $[Ca^{2+}]$, an increasing number of laboratories are utilizing these indicators to monitor nuclear, sarcoplasmic reticulum (SR), or mitochondrial $[Ca^{2+}]$ as well.

Mitochondria are the only organelles that are known to have a significant membrane potential with negative charge inside. Consequently, lipophilic compounds with a net positive charge are taken up by mitochondria to a much greater extent than by other cellular organelles (Chen 1989). Thus, it was rationalized that a fluorescent indicator for Ca^{2+} with a net positive charge would be preferentially compartmentalized in mitochondria, thereby providing a means to monitor the matrix $[Ca^{2+}]$. One such positively charged compound is the AM ester of Rhod-2 which was developed in 1989 by Roger Tsien and colleagues (Minta et al. 1989).

Rhod-2 has been used successfully by a number of laboratories to monitor changes in the mitochondrial matrix $[Ca^{2+}]$ in a wide variety of cell types including: adrenal chromaffin cells (Babcock et al. 1997); astrocytes (Jou et al. 1996; Simpson et al. 1998); cardiac myocytes (Sheu and Jou 1994; Trollinger et al. 1997); hepatocytes (Hajnóczky et al. 1995); lymphocytes (Hoth et al. 1997); neurons (Peng and Greenamyre 1998); oligodendrocytes (Simpson and Russell 1996); and smooth muscle cells (Mix et al. 1994). In this section, the properties of Rhod-2, and procedures for loading it into mitochondria, will be described. Following this, some results obtained with Rhod-2 as a monitor of mitochondrial [Ca^{2+}] in smooth muscle cells will be presented and then a consideration of some problems likely to be encountered by the experimenter using Rhod-2. For those who have no prior experience using Ca^{2+}-sensitive fluorescent indicators, excellent introductory material is available (Tsien 1989; Kao 1994; Haugland 1996).

Materials

The chemical structure of Rhod-2 is shown in Figure 2. The indicator comprises the highly fluorescent xanthene chromophore rhodamine coupled to the Ca^{2+} specific chelator, BAPTA (Minta et al. 1989). In the presence of Ca^{2+} the fluorescence maximum for excitation is 552 nm and for emission it is 581 nm. Rhod-2 has an effective K_d for Ca^{2+} of 570 nM (Haugland, 1996) and upon binding Ca^{2+} its fluorescence emission increases more than 100 fold (Figure 3). Rhod-2 is a single-wavelength, intensity-modulating indicator which means that changes in [Ca^{2+}] bring about changes in the intensity of its excitation and emission spectra, whereas the spectral maxima remain essentially unchanged. In contrast, ratiometric indicators (like Fura-2 and Indo-1) show not only intensity changes with changing [Ca^{2+}] but the Ca^{2+}-free and Ca^{2+}-bound forms of the indicator have distinct spectra. Therefore, intensity measurements can be made at two different wavelengths and a ratio obtained (Grynkiewicz et al. 1985). Rationing minimizes the effect of many artifacts that are unrelated to changes in [Ca^{2+}] and this will be discussed in later sections.

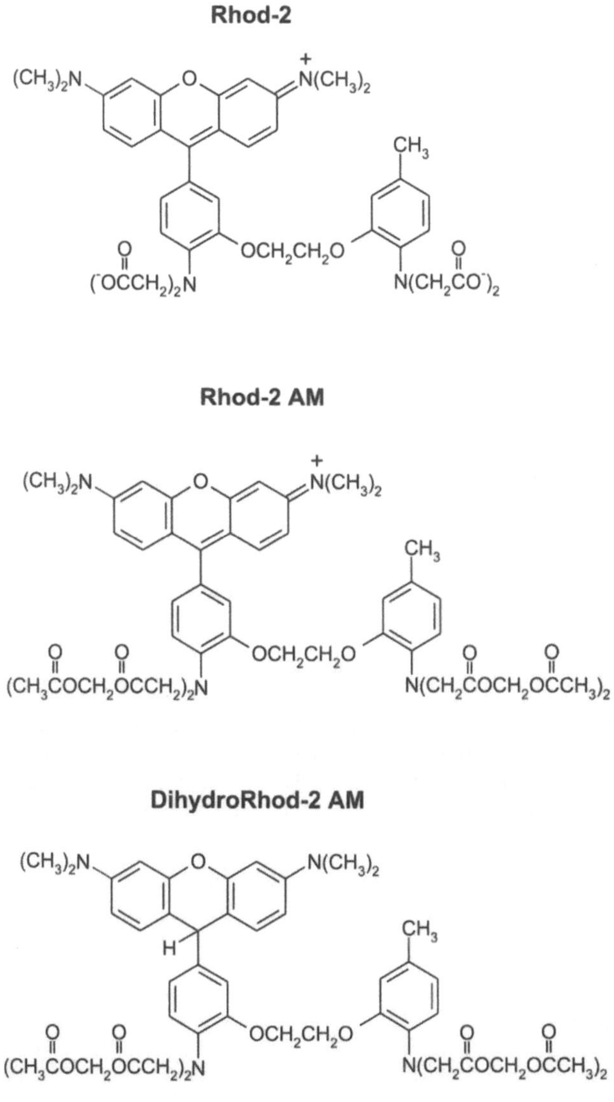

Fig. 2. Chemical structures of: Rhod-2 which is Ca^{2+}-sensitive (top); Rhod-2 AM which is Ca^{2+}-insensitive and membrane permeable (middle); and dihydroRhod-2 which is non-fluorescent and membrane permeable (bottom).

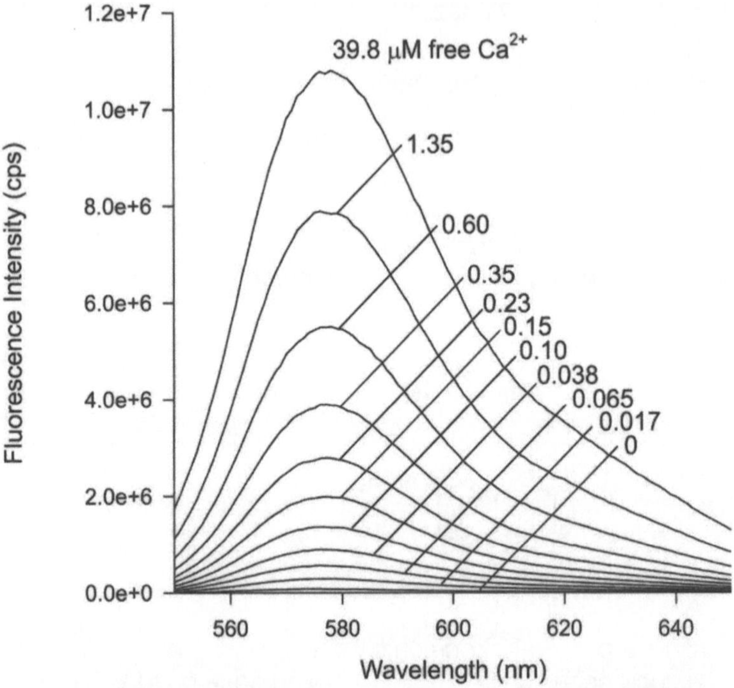

Fig. 3. Emission spectra for Rhod-2. 10 μM Rhod 2 was excited at 540 nm in a solution comprising 100 mM KCl, 10 mM MOPS (3-[N-Morpholino]propane-sulfonic acid) and the indicated free [Ca^{2+}]. The free [Ca^{2+}] was achieved by mixing appropriate amounts of 10 mM K_2EGTA, and 10 mM CaEGTA solutions.

Subprotocol 1

Procedure

Rhod-2 Loading into Mitochondria

The most widely used fluorescent indicators for Ca^{2+} are polycarboxylate anions, which cannot cross lipid bilayer membranes and are therefore cell impermeant. In order to make them membrane permeable, AM esters of these indicators were developed which mask the negative charges on the carboxyl group (Tsien et al. 1982). Subsequent incubation of cells, or tissue, in a dilute solution of the AM ester form of the indicator leads to cellular loading. Once inside the cell, cellular esterases hydrolyse the AM group, leaving the Ca^{2+}-sensitive polycarboxylate form trapped inside the cells. This trapping of the indicator allows cells to accumulate concentrations of up to hundreds of micromolar of the Ca^{2+}-sensitive form of these molecules, when incubated with 1-10 μM of the AM ester in the extracellular medium.

The AM ester of Rhod-2 is the only cell-permeant Ca^{2+} indicator which has a net positive charge (Fig. 2). This property readily promotes compartmentalization of Rhod-2 AM into mitochondria, most likely via mitochondrial membrane potential driven uptake. Once inside the mitochondrial matrix, esterases hydrolyse the AM group leaving Rhod-2 trapped in the matrix by virtue of its 3 negative charges and membrane impermeability.

Rhod-2 AM is normally prepared as a stock solution in anhydrous DMSO, at concentrations ranging from 0.1 to 10 mM. Aliquots of such stock solutions can be stored in the dark, desiccated at -20 °C for many months without apparent degradation. The nonionic and nondenaturing detergent Pluronic F-127 is frequently added to help disperse AM esters, prior to addition in the loading medium (Kao 1994). This can be achieved by mixing the aliquot of AM ester stock with an equal volume of 20% (w/v) Pluronic F-127, before being added into the aqueous loading medium. The final Pluronic concentration should not exceed 0.02%. To load the mitochondria with Rhod-2, cells are typically incubated with Rhod-2 AM (1-10 μM), for time periods which have ranged from 5 to 60 minutes. Although this concentration range should provide a good starting point for developing optimal mitochondrial loading conditions, obviously a certain amount of trial and error will be involved initially. In general, it is desirable to use the minimum indicator concentration to yield adequate fluorescence signal to noise levels.

Temperature

While the entry of Rhod-2 AM into cells occurs through the process of diffusion, the hydrolysis of the AM group is a chemical process requiring cellular esterases. Thus, removal of the AM group is likely to be more sensitive to changes in temperature than diffusion. In agreement with this notion, it has recently been reported that loading cells with Rhod-2 AM at low temperatures (4 °C) enhances the accumulation of Rhod-2 inside mitochondria relative to the cytosol (Trollinger et al. 1997). However, in the majority of studies using Rhod-2 to monitor mitochondrial [Ca^{2+}], loading has generally been carried out at room temperature.

Normally after Rhod-2 AM loading, the cells are maintained for a further period in indicator free medium for times ranging from 0.5-3 hours. This allows the cellular esterases sufficient time to remove the AM group and generate the Ca^{2+}-sensitive indicator. In studies where temperature was reported, this post-loading period is often carried out at higher temperatures (up to 37 °C). The reason for this is that at higher temperatures, cytosolic indicator is lost more rapidly than indicator in organelles. Consequently, compartmentalized indicator becomes a larger proportion of the total cellular indicator content (Kao 1994). It has also been reported that a much longer post-loading period (5-24 hours) in indicator free medium results in significant loss of Rhod-2 from the cytosol, while that already trapped in the mitochondrial matrix is retained (Hajnóczky et al. 1995; Simpson and Russell 1996; Trollinger et al. 1997).

DihydroRhod-2 AM

It is known that inside living cells, the colorless dihydrorhodamines and dihydrotetramethylrosamine are oxidized to fluorescent products that stain mitochondria (Chen 1989). This prompted some laboratories to examine whether the reduced form of Rhod-2 AM, dihydroRhod-2 AM (Fig. 2), would also enhance the compartmentalization of this indicator in mitochondria. Indeed, several groups have reported that by reducing Rhod-2 AM to the colourless, nonfluorescent dihydroRhod-2 AM, the discrimination between cytosolic and mitochondrially located indicator is improved (Hajnóczky et al. 1995; Bowser et al. 1998).

Preparation of dihydroRhod-2 is relatively straightforward procedure and can be carried out using the following protocol: Prepare a 1 mM stock solution of Rhod-2 AM in anhydrous DMSO. The reducing agent, sodium borohydride ($NaBH_4$) is prepared at a concentration of 1 mg/ml in methanol. Gradually add the minimum volume of $NaBH_4$ (typically 3-5 μl) to the stock Rhod-2 AM solution, such that the Rhod-2 AM just turns colorless. Thereafter, dihydroRhod-2 is loaded into the cells according to the same procedure described above for Rhod-2 AM. DihydroRhod-2 AM exhibits Ca^{2+}-dependent fluorescence only after it has been oxidized and its AM esters are cleaved to yield the indicator Rhod-2. It should be noted that dihydroRhod-2 AM will spontaneously, and quite rapidly, revert to the oxidized form characterized by the reappearance of color in the stock solution. Therefore, it should be used fairly soon after its preparation.

Calibration

For single wavelength indicators like Rhod-2, the free $[Ca^{2+}]$ is given by the equation: $[Ca^{2+}] = K_d * ((F-F_{min})/(F_{max}-F))$, where F_{min} is the indicator fluorescence intensity in the absence of Ca^{2+}, F_{max} is the indicator fluorescence intensity in the presence of saturating Ca^{2+}, and F is the measured fluorescence intensity for which $[Ca^{2+}]$ is to be calculated. In order to determine the $[Ca^{2+}]$ from the measured fluorescence (F), K_d, F_{min} and F_{max} all must be known. For Ca^{2+} indicators which are present in the cytosol, rough estimates of F_{min} and F_{max} are obtained by treating cells with media containing ionophores (e.g., ionomycin or A23187, which increase the permeability of the plasma membrane to divalent cations), in the presence of high concentrations of EGTA (F_{min}), or excess Ca^{2+} (F_{max}).

For the mitochondrially located Rhod-2, achieving accurate measurements for F_{min} and F_{max} is even more challenging. In this case the controlled external Ca^{2+} concentration has to be equilibrated with the mitochondrial matrix Ca^{2+} concentration. Only one group has attempted to calibrate the mitochondrial Rhod-2 fluorescence signal using the above approach; and they reported that depolarization of chromaffin cells increases mitochondrial Ca^{2+} concentration from ~200 nM to 600 nM (Babcock et al. 1997). However, in the majority of studies using single-wavelength intensity modulating indicators, fluorescence intensity is not converted to $[Ca^{2+}]$. Rather data are reported as either F/F_{rest} or DF/F_{rest}, where F is the fluorescence, DF is the change in fluorescence, and F_{rest} is resting (pre-stimulation) fluorescence.

Results

Rhod-2 has been used to monitor mitochondrial $[Ca^{2+}]$ in smooth muscle cells, both at the single cell level and also at the level of individual mitochondria in situ (Mix et al. 1994).

Figure 4 shows a smooth muscle cell, isolated from the stomach of the toad *Bufo marinus* (Fay et al. 1982), in which the mitochondria have been labeled with Rhod-2. In this instance, the loading protocol involved incubating the smooth muscle cells for 1 hour at room temperature with 1 μM Rhod-2 AM, followed by a post-incubation period (1-2 hours) in indicator free medium. The mitochondria are clearly identified by virtue of their characteristic rod-like appearance.

Figure 5 shows an experimental record where cytosolic and mitochondrial [Ca^{2+}] were simultaneously monitored in a rat pulmonary arterial smooth muscle cell. Smooth muscle cells were isolated as described and incubated with 2 μM Rhod-2 AM, for 1 hour at room temperature. This was followed by a 1 hour post-loading period in indicator free medium. Cytosolic [Ca^{2+}] was monitored with Fura-2 (50 μM), which was in-

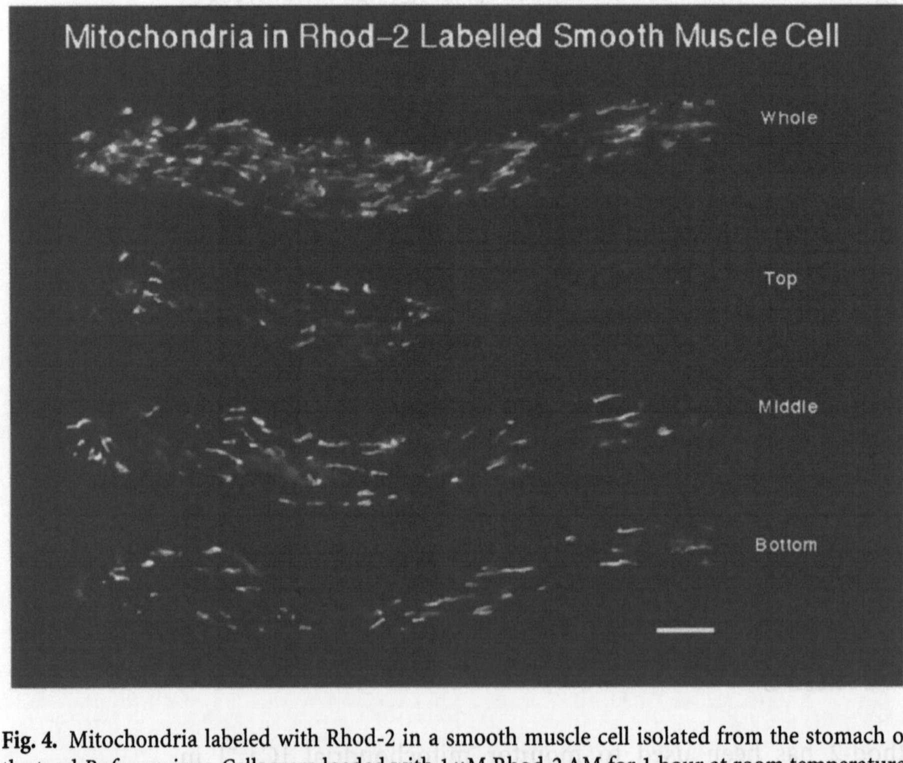

Fig. 4. Mitochondria labeled with Rhod-2 in a smooth muscle cell isolated from the stomach of the toad *Bufo marinus*. Cells were loaded with 1 μM Rhod-2 AM for 1 hour at room temperature. The uppermost image shows a projection of the entire 3D data set, while beneath single optical sections near the top, middle and bottom of the cell are shown. Scale bar is 10 μm.

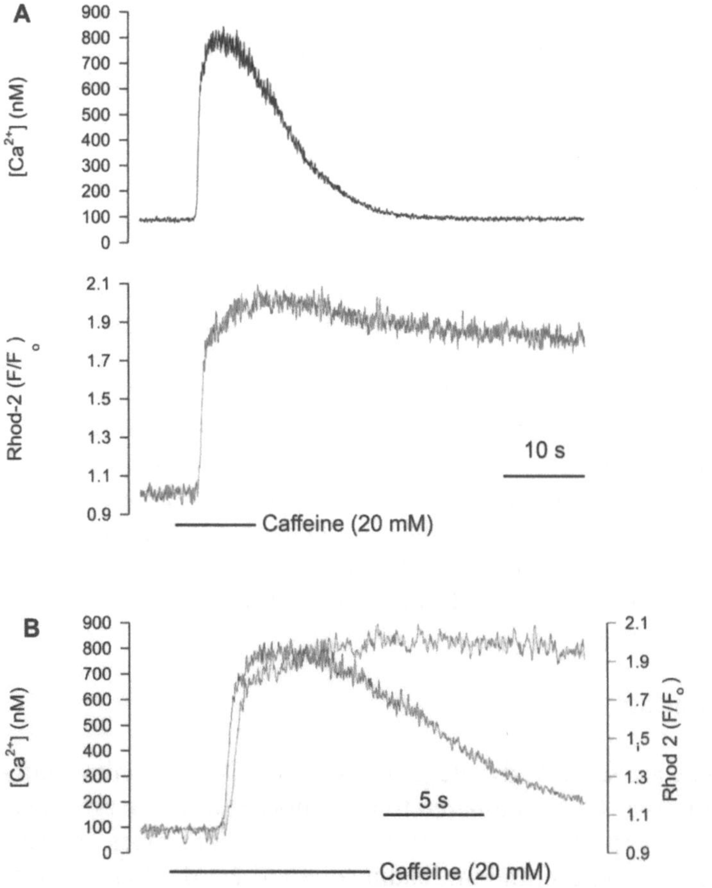

Fig. 5. Simultaneous measurement of cytosolic and mitochondrial [Ca^{2+}] in rat pulmonary artery smooth muscle cell. Cells were loaded with 2 μM Rhod-2 AM for 1 hour at room temperature in order to monitor mitochondrial [Ca^{2+}], and 50 μM Fura-2 was included in the patch pipette to enable measurement of cytosol [Ca^{2+}]. Application of caffeine (20 mM) activates the ryanodine receptor and causes release of Ca^{2+} from the SR. This leads to an increase in cytosolic and mitochondrial [Ca^{2+}] (**A**). The excitation filters used were 340 and 380 nm (~10 nm bandwidth) for Fura-2 and 500/10nm for Rhod-2. A single 80 nm bandwidth, 560 nm center wavelength emission filter was used for both Fura-2 and Rhod-2 emission. When the change in [Ca^{2+}] is plotted on an expanded time scale, it can be seen that the increase in mitochondrial [Ca^{2+}] occurs after a delay with respect to the increase in cytosolic [Ca^{2+}] (**B**).

troduced into the cytosol by inclusion in the patch pipette solution. Following an 8 minute dialysis period, the cell was stimulated by a 10 s application of caffeine (20 mM) applied from a pressure ejection pipette positioned close to the cell. Caffeine activates ryanodine receptors, causing the release of Ca^{2+} from the SR. This results in an increase in both cytosolic and mitochondrial $[Ca^{2+}]$ (Fig. 5A). It is clear that the cytosolic $[Ca^{2+}]$ reaches its peak value before that of the mitochondria. Furthermore, cytosolic $[Ca^{2+}]$ is restored to resting levels while mitochondrial $[Ca^{2+}]$ remains elevated. When cytosolic and mitochondrial $[Ca^{2+}]$ are plotted on an expanded time scale (Fig. 5B), a lag between the increase in cytosolic and mitochondrial $[Ca^{2+}]$ can be appreciated. This approach of using a Ca^{2+}-sensitive fluorescent indicator confined to the cytosol while simultaneously monitoring mitochondrial $[Ca^{2+}]$ allows detailed studies of the kinetics of the $[Ca^{2+}]$ changes in these two compartments. A similar approach has also been used to simultaneously monitor the cytosolic and SR $[Ca^{2+}]$.

Advantages

There are several advantages in using Rhod-2 to monitor mitochondrial $[Ca^{2+}]$ compared to the mitochondrially targeted aequorin approach. First, AM esters of Ca^{2+} indicators are relatively easy to use, and the loading of Rhod-2 AM into mitochondria is more straight forward than the molecular biology techniques associated with targeting aequorin to mitochondria. In addition, the latter approach requires that the cells be cultured to enable transfection. Second, with Rhod-2 it is possible to monitor the mitochondrial $[Ca^{2+}]$ not only in single cells, but also at the level of individual mitochondria in situ (Mix et al. 1994). This latter feature is particularly important, given that mitochondria are reported to represent a heterogeneous population (Rizzuto et al. 1994; Rutter et al. 1996). Third, unlike aequorin which is consumed upon Ca^{2+} binding, and therefore can no longer report $[Ca^{2+}]$, the fluorescence process is cyclical. This means that the same fluorophore can be repeatedly excited and detected, unless the fluorophore is irreversibly destroyed in the excited state (a phenomenon known as photobleaching). Consequently, fluorescent indicators can be used to monitor $[Ca^{2+}]$ over much longer time periods.

Troubleshooting

The major disadvantage with the approach of using AM esters to load intracellular compartments is that the experimenter has very little control as to where the indicator eventually becomes compartmentalized. Although with Rhod-2 AM there is some justification for using a positively charged indicator to enhance mitochondrial compartmentalization, care must be taken to ensure that the fluorescence signal does indeed originate from the mitochondria.

Before Rhod-2 AM can be compartmentalized in the mitochondrial matrix, it has to pass through the cytosol, then into the mitochondria before the AM group is removed. Thus, it is highly probable that the cytosol is going to contain some Rhod-2 as well. Consequently, this cytosolic Rhod-2 signal will contaminate the mitochondrial Rhod-2 signal and produce inaccurate measurements of changes in mitochondrial [Ca^{2+}]. While the loading procedures described above should help minimize contamination by cytosolic Rhod-2, an additional step which has been used to eliminate any residual Rhod-2 from the cytosol, is to dialyze the cytosol using a patch pipette (Mix et al. 1994; Babcock et al. 1997; David et al. 1998). Clearly, this approach is best suited to laboratories set up to carry out simultaneous microfluorimetric and electrophysiological studies.

By using a Ca^{2+} indicator which is confined exclusively to the cytosol, while simultaneously monitoring mitochondrial [Ca^{2+}] with Rhod-2, it is possible to simultaneously obtain kinetic information pertaining to the time course of the Ca^{2+} signals in both these compartments (see Figure 5; also Hajnóczky et al. 1995, Babcock et al. 1997, Robb-Gaspers et al. 1998).

It is also imperative that appropriate control studies are carried out to confirm that the Rhod-2 signal originates from the mitochondria. Since mitochondrial Ca^{2+} uptake is dependent upon the mitochondrial potential gradient, collapsing this potential will prevent the mitochondria from taking up Ca^{2+}. Thus, it is expected that the Rhod-2 signal should be blocked by exposing the cells to the protonophore FCCP, which collapses the mitochondrial potential gradient. Ruthenium red, which inhibits the mitochondrial Ca^{2+} uniporter, should also prevent any increase in Rhod-2 fluorescence upon cell stimulation. Unfortunately, RR is not membrane permeable and has to be introduced into cells by microinjection or inclusion in a patch pipette.

As mentioned earlier, Rhod-2 is not a ratiometric Ca^{2+} indicator. This means that the fluorescence intensity monitored at one wavelength is the only experimental measurement that is related to $[Ca^{2+}]$. Consequently, intensity changes arising from factors unrelated to changes in $[Ca^{2+}]$ can confound the interpretation of fluorescence data. During an experiment potential errors can arise from changes in cell thickness (this is especially problematical in studies on contractile cells) and loss of indicator fluorescence either because of photobleaching or actual loss from the cell by leakage. Thus, appropriate care must be taken to ensure that changes in Rhod-2 fluorescence actually reflect changes in $[Ca^{2+}]$. Further problems in calibrating the signal of rhod-2 originate from partial hydrolysis of the AM esters and, at the single organelle level, by movements of the mitochondria themselves.

Mitochondrial calcium measurements using aequorin

Aequorin is a Ca^{2+}-sensitive photoprotein naturally present in the medusa *Aequorea victoria*. Its general properties, with particular incidence on those that enable it to be used as a calcium sensor, are extensively discussed in the chapter by Brini and colleagues, where the measurement of Ca^{2+} in the nucleoplasm using aequorin chimeras is presented. In this section we will concentrate on its use for specifically measuring mitochondrial Ca^{2+} levels in living mammalian cells.

Targeting strategies

Measurement of $[Ca^{2+}]$ in mitochondria requires the presence of aequorin in this compartment. The strategy followed to achieve the correct delivery of aequorin to mitochodria took advantage of the natural signal peptides present in nuclearly-encoded mitochondrial proteins. The vast majority of mitochondrial proteins are encoded in nuclear DNA, synthesized in the cytosol, and imported into the organelle. Their correct targeting is due to a signal peptide, usually present at the N-terminus. Different proteins possess different signal peptides, with no consensus sequence, which simply bear some common characteristics regarding their charge distribution (Gavel and von Heijne 1990).

Mitochondrial matrix

Subunit VIII of human cytochrome c oxidase possesses a 25 amino acid signal peptide at its N-terminus, which is cleaved by a matrix protease upon import; it is presumed that the action of the protease is dependent on a motif present within the first few amino acids of the mature protein (Rizzuto et al. 1989). We fused the initial 31 amino acids of this subunit, comprising the signal peptide and the first 6 amino acids of the mature protein, to the N-terminus of aequorin, thus achieving the correct delivery of functional aequorin to mitochondria (Rizzuto et al. 1992).

Mitochondrial intermembrane space

For the delivery of aequorin to the mitochondrial intermembrane space (MIMS), we exploited the characteristics of another mitochondrial protein. Glycerol phosphate dehydrogenase (GPD) is an enzyme present in the mitochondrial inner membrane with a C-terminal domain protruding into the MIMS. To target aequorin to this space, we fused the photoprotein to the C-terminal portion of GPD, thus maintaining aequorin's C-terminus unaltered, since it is essential for its luminescent properties (Rizzuto et al. 1998).

Subprotocol 2

Procedure

Many of the procedures used in the studies presented in this section are described in detail in another chapter. To avoid unnecessary repetition, only brief descriptions are given below; please refer to to the chapter "Measuring Ca^{2+} in the nucleoplasm of intact cells" by Brini et al. for details.

Growth of cells for $[Ca^{2+}]$ measurements

HeLa cells were used to measure matrix and MIMS $[Ca^{2+}]$ ($[Ca^{2+}]_m$ and $[Ca^{2+}]_{mims}$, respectively). The cells were seeded in coverglasses (Ø 13 mm) and allowed to grow until about 50% confluency. At this point the cells were transfected with 4 µg of the appropriate plasmid, using the calcium phosphate procedure, as explained in detail in the previous chapter. After 36 hours, aequorin was reconstituted by adding the prosthetic group to the incubation medium (for details see previous chapter). After the reconstitution (approx. 2 hours), the coverglass was inserted into the measuring system.

Treatment of cells with different agents

For the increase of $[Ca^{2+}]_c$ we used an agonist - histamine - that, acting on a G protein coupled-receptor, leads to the generation of inositol 1,4,5 trisphosphate (IP_3). Briefly, histamine (100 µM, final concentration) was added to the perfusion medium (modified KRB buffer), for 1 minute, followed by rinsing in the same buffer.

To collapse the electrochemical gradient across the mitochondrial inner membrane, the protonophore FCCP (2 µM, final concentration) was added to the perfusion medium.

Ruthenium red, a specific blocker of the mitochondrial Ca^{2+} uniporter, does not freely cross the cell membrane. Experiments using this drug (4µM, final concentration), were carried out in permeabilized cells.

To permeabilize cells, they were incubated for 1 min with digitonin (100 µM) added to an "intracellular buffer" (IB; 140 mM KCl, 5 mM NaCl, 1 mM K_3PO_4, 5.5 mM glucose, 2 mM $MgSO_4$, 1 mM ATP, 2 mM Na-succinate, 20 mM HEPES, pH 7.05 at 37 °C), supplemented with EGTA (100 µM), then washed in IB, and maintained in IB with a free $[Ca^{2+}]$ of 2 µM (buffered with 2 mM EGTA). All additions were made to this buffer.

Immunolocalization

HeLa cells were seeded in coverglasses (Ø 24 mm) and allowed to grow until about 50% confluency. At this point the cells were transfected with 8 µg of the appropriate plasmid, using the calcium phosphate procedure. After 36 hours the cells were processed for immunofluorescence as follows. The cells were fixed with 3.7% formaldehyde in PBS for 20 min, washed three times with PBS and then incubated for 10 min in PBS supplemented with 50 mM NH_4Cl. Permeabilization of cell membranes was obtained by incubation with 0.1% Triton X-100 in PBS for 5 min, followed by a 1 hour wash with 1% gelatin (type IV, from calf skin) in PBS. The cells were then incubated for 1h at 37 °C in a wet chamber with a 1:200 dilution (in PBS) of the monoclonal antibody 12CA5, which recognizes the HA1 epitope (see below).

Staining was then carried out with TRITC-labelled anti-mouse secondary antibodies. After each antibody incubation, the cells were washed four times with PBS. Images were obtained with a Zeiss Axioplan equipped with a Princeton Instruments back-illuminated cooled CCD camera, and analysed with Universal Imaging's Metamorph software package.

Results

Subcellular fractionation

Demonstrating the correct localization of aequorin chimeras. Cells grown to confluence in large (Ø 15 cm) Petri dishes were scraped into 5 ml of PBS, washed, and ruptured by repeated passages through a 22G needle. This suspension was centrifuged to remove nuclei and unbroken cells (pellet). The supernatant was centrifuged to pellet mitochondria (Fraction I); the accompanying supernatant was re-centrifuged to pellet microsomes (Fraction II), while the resulting supernatant is constituted essentially by cytosolic elements (Fraction III). On each of these fractions the activities of lactate dehydrogenase, citrate synthase, and aequorin were measured. The fractionation studies demonstrate the correct mitochondrial localization of mtAEQ (Table 1).

Immunolocalization

To confirm the correct localization of the aequorin chimeras, we performed the following studies. In all constructs made, we added a short amino acid sequence (HA1, derived from the influenza virus hemagglutinin protein) to the C-terminus of aequorin. This epitope can be readily detected by antibodies de-

Table 1. Subcellular distribution of marker enzymes and mitochondrially-targeted aequorin. Values represent percentage of total enzyme (for LDH and CS) or aequorin activity. (Adapted from Rizzuto et al. 1994).

	Fraction I (mitochondria)	Fraction II (microsomes)	Fraction III (cytosol)
Lactate dehydrogenase (LDH)	5	5	86
Citrate synthase (CS)	98.5	0.5	1
Aequorin	97.5	0.5	2

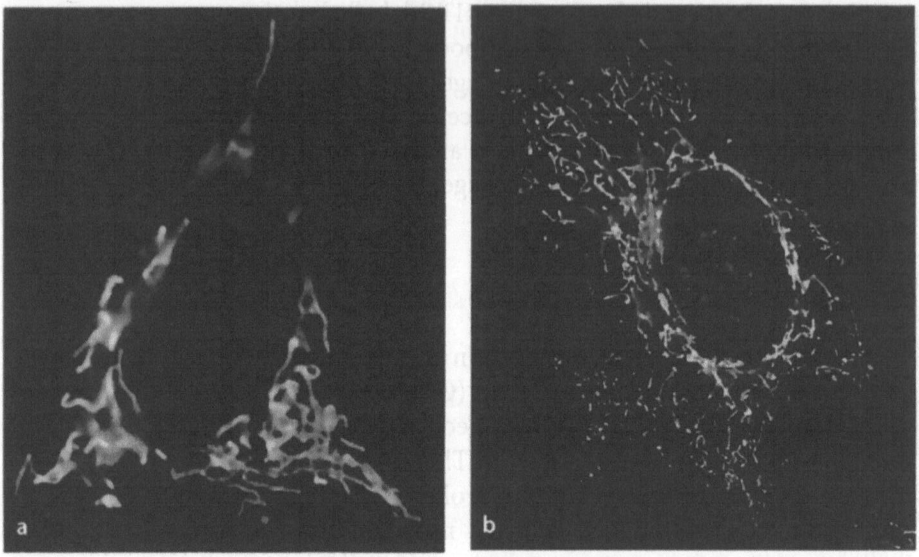

Fig. 6. Immunolocalization of mtAEQ (a) and mimsAEQ (b) in HeLa cells.

signed against the nonapeptide sequence YPYDVPDYA. Cells transiently expressing aequorin targetted to the mitochondrial matrix (mtAEQ) or the mitochondrial intermembrane space (mimsAEQ) exhibit the patterns shown in Figure 6, typical of a mitochondrial network; at this level of resolution, the two chimeras are not distinguishable, but their different localization is investigated below.

Large mitochondrial Ca^{2+} responses occur in most cell types

The biphasic kinetics of $[Ca^{2+}]_c$ in HeLa cells transfected with wild-type aequorin (without any targeting sequences and thus cytosolic) is shown in the inset of Figure 7. The release of Ca^{2+}

Fig. 7. Agonist-dependent increases in mitochondrial Ca^{2+} concentrations in a variety of cell types. In all cases the reporter chimera is mtAEQ (except in the **inset** of **A**, where the cytoplasmic response is observed in HeLa cells transfected with cytAEQ); the cells, continuously perfused with KRB, were challenged with the indicated agonists. In **A** and **B**, the response of HeLa cells and fibroblasts (cell line 143B), respectively, upon stimulation with histamine; in **C**, the response of Chinese hamster ovary epithelial cells (CHO) to ATP; **D**, neurons stimulated by glutamate (Glu); **E**, primary cholangiocytes stimulated with ATP; **F**, myocytes stimulated with carbachol (Cch), in the presence of atropine. These experiments demonstrate the usefulness of mtAEQ in determining $[Ca^{2+}]_m$ in different cell types and in the presence of different agonists.

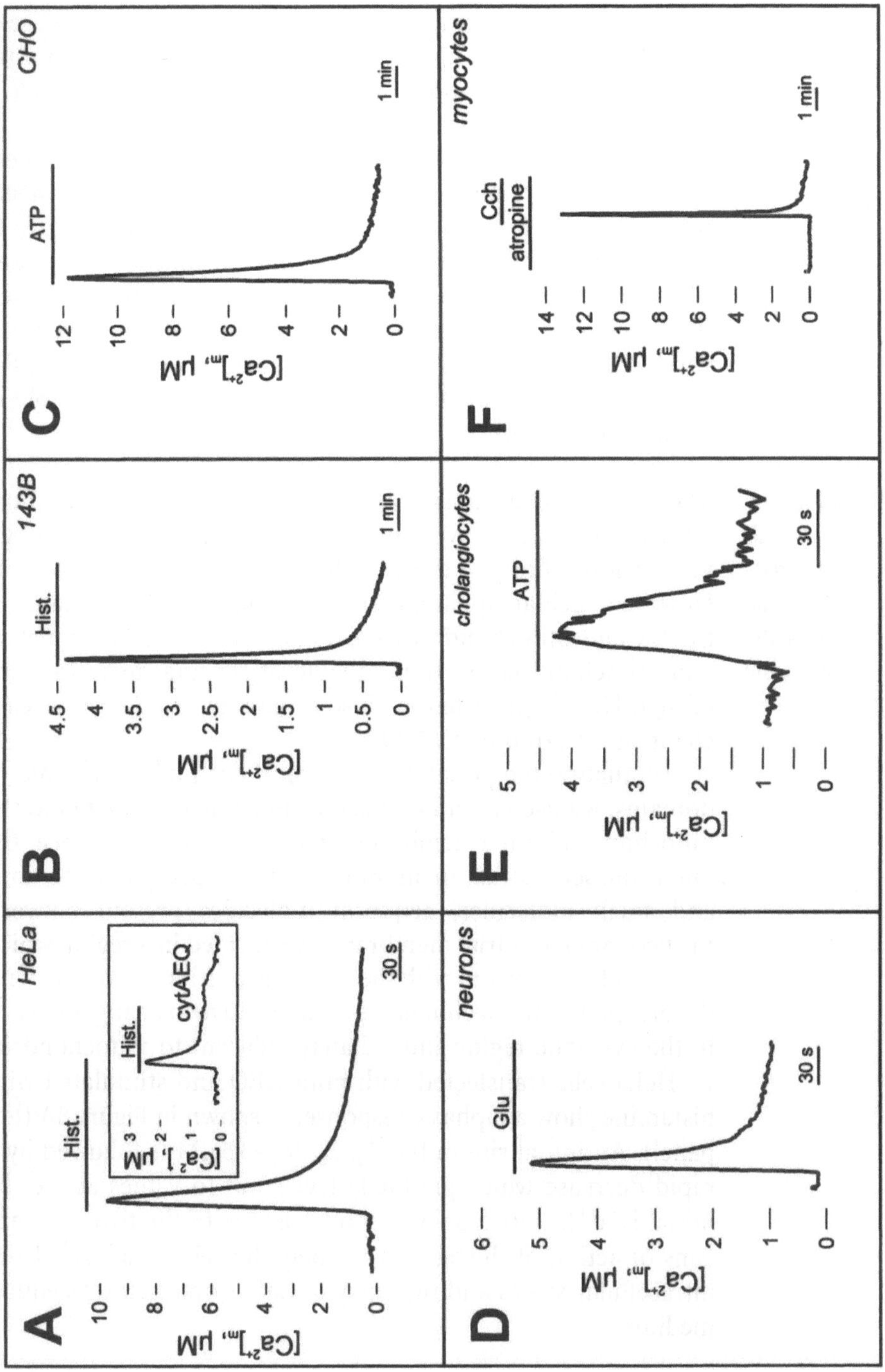
A
HeLa
Hist.
cytAEQ
30 s
B
143B
Hist.
1 min
C
CHO
ATP
1 min
D
neurons
Glu
30 s
E
cholangiocytes
ATP
30 s
F
myocytes
Cch
atropine
1 min
[Ca2+]m, μM

from the stores causes an initial rapid, but transient, increase of $[Ca^{2+}]_c$ (to values of approx. 2.5 μM), followed by a sustained increase of $[Ca^{2+}]_c$ above normal basal levels, and which is maintained throughout the stimulation period.

Figure 7A shows the measurements obtained in HeLa cells transiently expressing mtAEQ. Upon stimulation with histamine, there is a rapid increase in $[Ca^{2+}]_m$ up to approx. 10 μM, followed by the return, within 2 min, to almost basal values. The usefulness of this chimeric probe is exemplified in panels B-F of Figure 7, where large $[Ca^{2+}]_m$ changes are observed in a variety of cell lines, as well as different primary cultures. Similar studies can be conducted in virtually any cell type capable of being transfected.

The MIMS chimera reveals Ca^{2+} rises higher than those of the bulk cytosol

The fact that alterations in $[Ca^{2+}]_m$ are much larger than those observed for $[Ca^{2+}]_c$ were originally unexpected, especially given the low affinity of the mitochondrial uniporter for this ion. However, such an apparent contradiction could be explained if the ER and mitochondria have a close structural relationship, and mitochondria are thus capable of sensing microdomains of high $[Ca^{2+}]$ generated in close proximity of ER Ca^{2+} release channels (Rizzuto et al. 1998).

To analyse the possible existence of such high Ca^{2+} microdomains, we used a different aequorin chimera, targeted to the mitochondrial intermembrane space (mimsAEQ). Since the outer mitochondrial membrane is freely permeable to ions and small molecules, aequorin molecules present between the two mitochondrial membranes are located in a region which is in rapid equilibrium with the cytosolic portion in contact with the organelle. Such a chimera is thus sensitive to changes in Ca^{2+} in the cytosolic region immediately adjacent to mitochondria.

HeLa cells transfected with mimsAEQ and stimulated with histamine show a biphasic response, as shown in Figure 8A (left panel). An initial rise in $[Ca^{2+}]_{mims}$ to ~3.5 μM is followed by a rapid decrease which gradually levels out to values above the initial $[Ca^{2+}]$. This type of response is due to the two mechanisms of action of the agonist, namely the release of Ca^{2+} from intracellular stores and the entry of Ca^{2+} from the extracellular medium.

The comparison of $[Ca^{2+}]_c$ and $[Ca^{2+}]_{mims}$ responses to histamine shows a clear difference only in the initial phase (that which is due to the release of Ca^{2+} from intracellular stores). These data support the hypothesis that the opening of InsP3-sensitive channels in close proximity to mitochondria generate microdomains of high $[Ca^{2+}]$. Indeed, in such microdomains the $[Ca^{2+}]$ is much higher than the average $[Ca^{2+}]_c$. and thus the low affinity Ca^{2+} uptake systems present in mitochondria are capable of efficiently accumulating Ca^{2+} in the matrix of the organelle. Interested readers are referred to (Rizzuto et al. 1998), since the details of the experimental procedures regarding the testing of this hypothesis are clearly beyond the scope of this chapter.

The use of mitochondrial drugs to elucidate fine details

Given the targeting signal, it is expected that the chimeric protein mtAEQ is localized in the mitochondrial matrix. However, our immunofluorescence data (above) indicated only a general mitochondrial localization of mtAEQ, and it is not possible to discriminate its exact localization within the organelle. To ensure that our $[Ca^{2+}]$ measurements were indeed due to changes in the mitochondrial matrix, we analyzed the effect of uncouplers on the $[Ca^{2+}]$ values generated by our chimeric probe. Since Ca^{2+} entry into the mitochondrial matrix is made possible by the electrochemical gradient across the inner mitochondrial membrane, collapse of this gradient with an uncoupler should abolish the changes in $[Ca^{2+}]$ measured by our probe. Using FCCP, that dissipates the proton gradient across the IMM, we were unable to detect a significant response to the histamine stimulus, as reported in Figure 8B. Further confirmation that our probe is indeed localized in the mitochondrial matrix and that calcium accumulation occurs via the known uptake mechanism (the Ca^{2+} uniporter) came from experiments with a specific inhibitor, Ruthenium red. Given that cells are not permeable to RR, cells were first permeabilized with digitonin in EGTA-containing IB buffer, as described in the procedures; calcium uptake was initiated by supplementing the medium with known concentrations of Ca^{2+}. It is apparent from Figure 8C that upon challenging with 2 μM free Ca^{2+}, $[Ca^{2+}]_m$ increases to over 3 μM; in the presence of RR, uptake was almost completely abolished.

To determine the exact location of mimsAEQ we performed the following experiments. We had previously determined that the collapse of the proton gradient drastically reduces the accumulation of Ca^{2+} in the matrix. If mimsAEQ were localized within the matrix, the rise in $[Ca^{2+}]$ should be abolished in the presence of uncouplers. Conversely, if mimsAEQ were localized in the space between the two mitochondrial membranes, then the presence of uncouplers should not affect the $[Ca^{2+}]$ values obtained. As shown in Figure 8A (right panel), HeLa cells transfected with mimsAEQ and treated with FCCP do not show a significant difference in $[Ca^{2+}]$ dynamics, when compared to similar, coupled cells, strongly suggesting that the aequorin moiety lies in the MIMS.

Taken as a whole, these experiments demonstrate the usefulness of aequorin as a calcium probe in a wide range of circumstances. Aequorin can be targeted to different subcellular locations (including specific organelle compartments), where it can sense specific calcium changes. Moreover, the use of different mitochondrial drugs can be used to analyse calcium changes in a variety of conditions, as exemplified by – but not limited to – the cases documented above.

Fig. 8. Effect of different drugs on the matrix and MIMS $[Ca^{2+}]$ of HeLa cells. A. Variation of $[Ca^{2+}]_{mims}$ upon stimulation with histamine, in the absence and presence of FCCP (left and right panels, respectively). **B.** variation of $[Ca^{2+}]_m$ upon stimulation with histamine, in the absence and presence of FCCP (left and right panels, respectively). C. variation of $[Ca^{2+}]_m$ in the absence (control) and in the presence (+RR) of Ruthenium red, upon stimulation with histamine in Ca^{2+}-containing medium *(b)*, after cell permealization and perfusion in Ca^{2+}-free medium *(a)*.

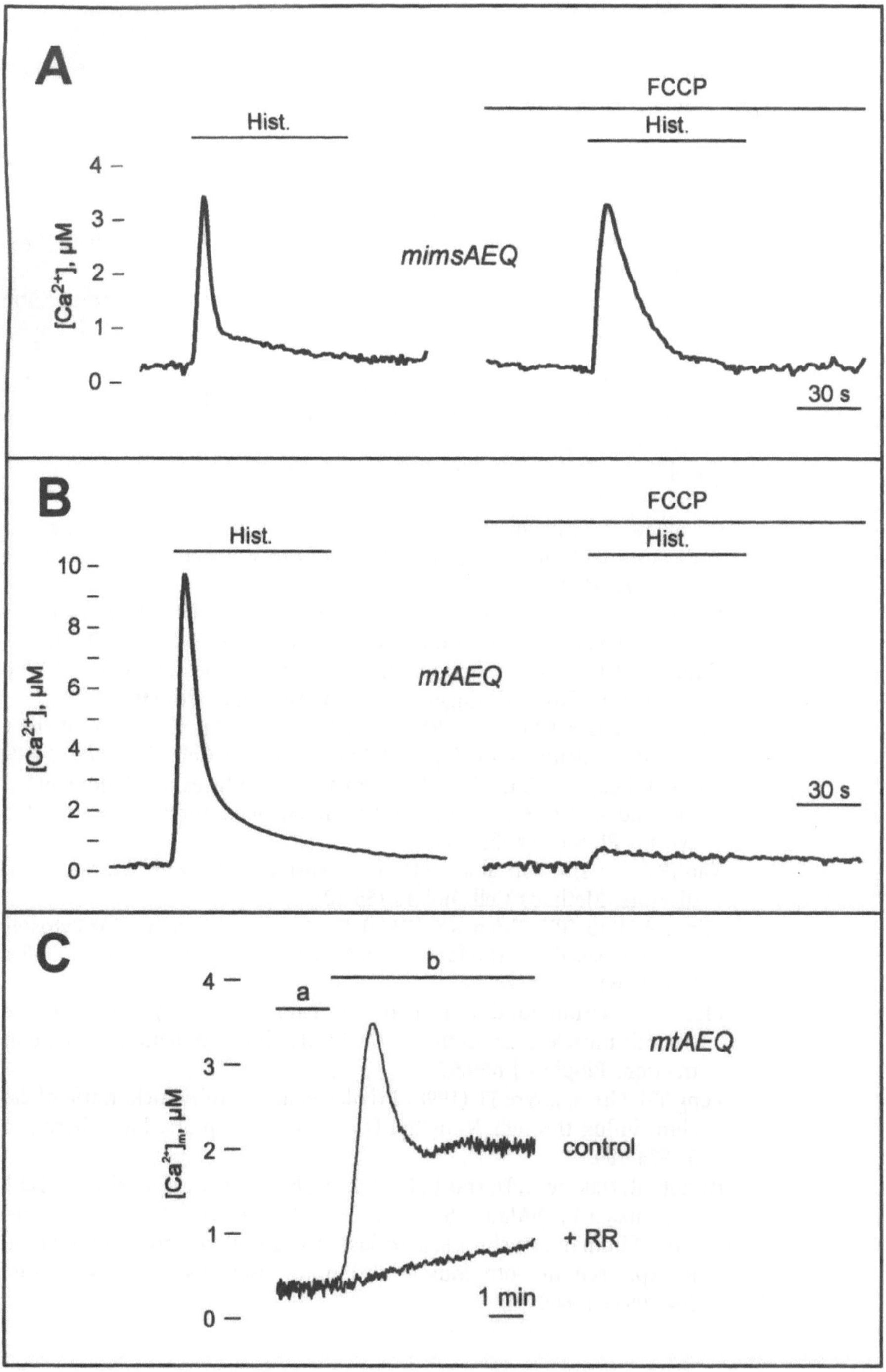
A
FCCP
Hist.
Hist.
4
3
2
1
0
[Ca2+], µM
mimsAEQ
30 s
B
FCCP
Hist.
Hist.
10
8
6
4
2
0
[Ca2+], µM
mtAEQ
30 s
C
b
a
4
3
2
1
0
[Ca2+]m, µM
mtAEQ
control
+ RR
1 min

References

Babcock DF, Herrington J, Goodwin PC, Park YB, Hille B (1997) Mitochondrial participation in the intracellular Ca^{2+} network. J Cell Biol 136:833-844.

Bowser DN, Minamikawa T, Nagley P, Williams DA (1998) Role of mitochondria in calcium regulation of spontaneously contracting cardiac cells. Biophys J 75:2004-2014.

Chen LB (1989) Fluorescent labeling of mitochondria. Methods Cell Biol 29:103-123.

David G, Barrett JN, Barrett EF (1998) Evidence that mitochondria buffer physiological Ca^{2+} loads in lizard motor nerve terminals. J Physiol 509:59-65.

Fay FS, Hoffman R, LeClair S, Merriam P (1982) Preparation of individual smooth muscle cells from the stomach of Bufo marinus. Methods Enzymol 85:284-292.

Gavel Y, von Heijne G (1990) Cleavage site motifs in mitochondrial targeting peptides. Protein Eng. 4:33-37.

Grynkiewicz G, Poenie M, Tsien RY (1985) A new generation of Ca^{2+} indicators with greatly improved fluorescence properties. J Biol Chem 260:3440-3450.

Hajnóczky G, Robb-Caspers RD, Seitz MB, Thomas AP (1995) Decoding of cytosolic calcium oscillations in the mitochondria. Cell 82:415-424.

Haugland RP (1996). Handbook of Fluorescent Probes and Research Chemicals (6th Edition). Eugene, Oregon, Molecular Probes.

Hoth M, Fanger CM, Lewis RS (1997) Mitochondrial regulation of store-operated calcium signaling in T lymphocytes. J Cell Biol 137:633-648.

Jou M-J, Peng T-I, Sheu S-S (1996) Histamine induces oscillations of mitochondrial free Ca^{2+} concentration in single cultured rat brain astrocytes. J Physiol 497:299-308.

Kao JPY (1994) Practical aspects of measuring [Ca^{2+}] with fluorescent indicators. Methods Cell Biol 40:155-181.

Minta A, Kao JPY, Tsien RY (1989) Fluorescent indicators for cytosolic calcium based on rhodamine and fluorescein chromophores. J Biol Chem 264:8171-8178.

Mix TCH, Drummond RM, Tuft RA, Fay FS (1994) Mitochondria in smooth muscle cells sequester Ca^{2+} following stimulation of cell contraction. Biophys J 66:A97.

Peng T-I, Greenamyre JT (1998) Privileged access to mitochondria of calcium influx through *N*-methyl-D-aspartate receptors. Mol Pharmacol 53:974-980.

Rizzuto R, Nakase H, Darras B, Francke U, Fabrizi GM, Mengel T, Walsh F, Kadenbach B, DiMauro S, Schon EA (1989) A gene specifying subunit VIII of human cytochrome c oxidase is localized to chromosome 11 and is expressed in both muscle and non-muscle tissues. J Biol Chem 264:10595-10600.

Rizzuto R, Pinton P, Carrington W, Fay FS, Fogarty KE, Lifshitz LM, Tuft RA, Pozzan T (1998) Close contacts with the endoplasmic reticulum as determinants of mitochondrial Ca^{2+} responses. Science 280:1763-1766.

Rizzuto R, Simpson AWM, Brini M, Pozzan T (1992) Rapid changes of mitochondrial Ca^{2+} revealed by specifically targeted recombinant aequorin. Nature 358:325-327.

Rizzuto RR, Bastianutto C, Brini M, Murgia M, Pozzan T (1994) Mitochondrial Ca^{2+} homeostasis in intact cells. J Cell Biol 126:1183-1194.

Robb-Gaspers LD, Rutter GA, Burnett P, Hajnóczky G, Denton RM, Thomas AP (1998) Coupling between cytosolic and mitochondrial calcium oscillations: role in the regulation of hepatic metabolism. Biochim Biophys Acta 1366:17-32.

Rutter GA, Burnett P, Rizzuto R, Brini M, Murgia M, Pozzan T, Tavaré JM, Denton RM (1996) Subcellular imaging of intramitochondrial Ca^{2+} with recombinant targeted aequorin: significance for the regulation of pyruvate dehydrogenase activity. Proc Natl Acad Sci USA 93:5489-5494.

Sheu S-S, Jou M-J (1994) Mitochondrial free Ca^{2+} concentration in living cells. J Bioenerg Biomembr 26:487-493.

Simpson PB, Mehorta S, Langley D, Sheppard CA, Russell JT (1998) Specialized distribution of mitochondria and endoplasmic reticulum proteins define Ca^{2+} wave amplification sites in cultured astrocytes. J Neurosci Res 52:672-683.

Simpson PB, Russell JT (1996) Mitochondrial support inositol 1,4,5-trisphosphate-mediated Ca^{2+} waves in cultured oligodendrocytes. J Biol Chem 271:33493-33501.

Trollinger DR, Cascio WE, Lemasters JJ (1997) Selective loading of Rhod 2 into mitochondria shows mitochondrial Ca^{2+} transients during the contractile cycle in adult rabbit cardiac myocytes. Biochem Biophys Res Commun 236:738-742.

Tsien RY (1989) Fluorescent indicators for ion concentrations. Methods Cell Biol 30:127-156.

Tsien RY, Pozzan T, Rink TJ (1982) Calcium homeostasis in intact lymphocytes: cytoplasmic free calcium monitored with a new, intracellularly trapped fluorescent indicator. J Cell Biol 94:325-334.

Suppliers

Most substances required to perform the techniques described are available from common suppliers of laboratory reagents. Anti-HA1 antibodies are available from BAbCo, and secondary antibodies from Dako; Rhod-2 AM and coelenterazine can be purchased from Molecular Probes.

Abbreviations

AM	acetoxymethyl
BAPTA	1,2-*bis*(*o*-aminophenoxy)ethane-N,N,N',N'-tetraacetic acid
DMSO	dimethylsulfoxide
FCCP	carbonyl cyanide *p*-(trifluoromethoxy)phenylhydrazone
RR	Ruthenium red
TRITC	tetramethylrhodamine isothiocyanate

Ca^{2+} Measurements in Secretory Granules

OLEG GERASIMENKO

Introduction

An important class of agonist-evoked cytosolic Ca^{2+}signals is due to messenger-mediated release of Ca^{2+} stored intracellularly. The most important calcium-releasing messenger is inositol 1,4,5-trisphosphate (IP_3) (Berridge 1993), although recently it has been demonstrated that cADP-ribose (Rakovic 1996) and NAADP (Cancela 1999) can also be important intracellular calcium-releasing messengers. It has been generally accepted, that agonists evoke cytosolic calcium signals via IP_3 - mediated calcium release from the endoplasmic reticulum (Berrige 1993, Clapham 1995).

In highly polarized pancreatic acinar cells, physiological concentrations of agonists like acetylcholine (ACh) or cholecystokinin (CCK) evoke cytosolic calcium spikes confined to the apical secretory pole, a region that mainly contains secretory granules. The intracellular infusion of IP_3 or cADP-ribose evokes spikes that are either confined to the secretory granular area or initiated in this part of the cell.

These spikes can occur alone or repetitively and in some cases precede longer lasting Ca^{2+} signals,which spread throughout the cell (Thorn 1993, Kasai 1993). The local Ca^{2+} spikes in the secretory pole are not due to regional CCK or ACh receptors in the apical part of the cell, since intracellular infusion of IP_3 (or a nonmetabolizable IP_3 analog) evokes repetitive Ca^{2+} spikes that are either confined to the secretory pole or initiated in

Oleg Gerasimenko, The University of Liverpool, The Physiological Laboratory, Crown Street, P.O. Box 147, Liverpool, L69 3BX, UK
(*phone* +44-151-794-5357; *fax* +44-151-794-5327;
e-mail olegl@liverpool.ac.uk)

this part of the cell before spreading (Thorn 1993). Selective micropipette injections of IP_3 into the secretory pole or the basal part of the cell reveal a much higher sensitivity of the Ca^{2+} release mechanism to the messenger in the secretory pole than in the basal part of the cell (Kasai 1993). It has also been shown that local ACh application at the base of the cell can evoke a cytosolic Ca^{2+} rise in the secretory pole, some considerable distance away from the site of messenger generation (Thorn 1993), most likely owing to rapid diffusion of the very mobile messenger IP_3 to the site with the highest sensitivity in the secretory pole (Kasai 1994). All these data indicate that in pancreatic acinar cells the secretory pole has a special role in the Ca^{2+} signaling because of its high sensitivity to IP_3 which is most likely either due to a relatively high concentration of IP_3 receptors in this region or to the particularly high affinity of the IP_3 receptors in the secretory pole for IP_3 (or to both causes) (Petersen 1994). An immunocytochemistry study has shown that the type 3 IP_3 receptor is specifically localized in the secretory pole of pancreatic acinar cells (Nathanson 1994). The local Ca^{2+} spikes in the secretory pole are of physiological importance, since they are sufficient to evoke spikes of exocytosis, as revealed by capacitance measurements (Maruyama 1993, 1994).

It is well known from numerous electron microscopic investigations that in pancreatic acinar cells the secretory pole is densely packed with zymogen granules (ZGs), whereas the nucleus and the ER dominate the basal part of the cell (Kern 1993, Gorelick 1994). The fact that the physiologically relevant cytosolic Ca^{2+} signals occur in the region mainly containing ZGs, and not in the part of the cell dominated by the ER, prompted us to investigate the possibility that ZGs, which are known to have a high Ca^{2+} content (Clemente 1975, Nakagaki 1984, Nicaise 1992) , may release part of their stored Ca^{2+} in response to IP_3 stimulation. Yoo and Albanesi (1990) described experiments in which IP_3 evoked an increase in the Ca^{2+} concentration (time to peak about 1 min) surrounding populations of isolated adrenal chromaffin granules, but this finding has remained controversial (Kasai 1995).

Using confocal microscopy it has been shown recently that changes of the free Ca^{2+} concentration occur inside single granules in intact cells or outside the granules (Nguyen 1998). The main approach that has so far given most significant data about calcium stores in secretory zymogen granules involves isolation of the granules (Yoo 1990), especially when the novel methods of calcium measurements on single isolated organelles were developed (Gerasimenko 1995) which made it possible to perform experiments with single isolated ZGs (Gerasimenko 1996).

Materials

- Laser scanning confocal microscope Odyssey (Noran Instruments) with 20 x and 100 x Nikon objectives, NA 1.3 oil, summing option 128 frames
- Mice male CD1
- Homogenizer Teflon/glass
- Refrigerated centrifuge for 2500 x g
- ATP, IP_3, 3-deoxy-3-fluoro IP_3, cADP-ribose, trypsin inhibitor, bovine serum albumin, α-amylase diagnostic kit, protein measuring kit, chemicals for the bath solution (Sigma)
- Fluo-3, Indo-1, Mag-fura red AM, DiOC6(3), LysoTracker, Calcium Green-1 dextran (70 kDa), and acridine orange (Molecular Probes)

Bath solution A
- 10 mM MOPS-Tris pH 6.8
- 0.1 mM $MgSO_4$
- 3 mM ATP
- 250 mM sucrose
- 0.5 mg/ml trypsin inhibitor
- 1 mg/ml bovine serum albumin

Bath solution B
- 135 mM KCl
- 10 mM MOPS-Tris pH 6.8
- 0.1 mM $MgSO_4$

- 3 mM ATP
- 0.1 mg/ml trypsin inhibitor
- 1 mg/ml bovine serum albumin

Procedure

Isolation of secretory granules

Note: All steps should be performed at 4 °C.

1. After killing 2-month-old mice by cervical dislocation rinse isolated pancreata with ice-cold bath solution A.
2. Chop finely isolated pancreata and homogenize in 3 ml of ice-cold buffer A using Teflon/glass homogenizer (8-10 strokes).
3. Mix with another 6-7 ml of ice-cold buffer A.
4. Centrifuge at 1000 x g for 10 min to remove nuclei and cellular debris.
5. Gently collect supernatant and centrifuge 2500 x g for 10 min.
6. Resuspend the resulting white pellet in buffer B.
7. Check the quality of isolated granules preparation by α-amilase assay using Sigma diagnostic kit and by measurement of total protein using the protein measuring kit (micro) from Sigma.

Loading of secretory granules with fluorescent indicators

1. For the purpose of staining, incubate the fraction of secretory granules with 1 µM Mag-fura red AM for 7 min at 20 °C.
2. Wash two times by centrifugation 2500 x g for 10 min in buffer B to rinse away extracellular Mag-fura red.

Ca^{2+} measurements

Note: All Ca measurements can be done at room temperature.

1. Place 10 μl of the preparation of secretory granules in the experimental chamber and add Fluo-3 or Indo-1 (pentapotassium salts) to obtain a 100 μM concentration of the fluorescent dye and Ca^{2+} buffer for measurements of the free Ca^{2+} concentration in the medium outside the granules.
2. Add IP_3 or cADP-ribose to attain final concentration 5–25 μM.
3. Calibrate the responses using $CaCl_2$ and EGTA at the end of experiments.
4. For measurements of changes in the free Ca^{2+} concentration in the medium surrounding isolated secretory granules, place 10 μl samples containing 100 μM Fluo-3 or Indo-1 (from 2 mM stocks in buffer B) in the experimental chamber on the stage of the confocal microscope with a 20 x Nikon objective.
5. Measurements with Fluo-3 should be done with an excitation wavelength of 488 nm and emission wavelengths of 510-560 nm. For Indo-1 use an excitation wavelength of 363 nm, and the ratios of emission at 405 nm and 495 nm.
6. For measurements of the Ca^{2+} concentration inside a single zymogen granule incubate secretory granules with 1 μM Mag-fura red AM for 7 min at 20 °C. After staining wash the granules by two subsequent centrifugations in buffer B.
7. Use Mag-fura red-loading granules (exitation 488 nm, emission ›590 nm) in suspension and single granule experiments. Dilute 10 μl of the preparation of secretory granules to a 100 μl volume with buffer B containing 100 μM EGTA and place it in the experimental chamber under confocal microscope.
8. Add or ionophoretically inject 5-25 μM of IP_3 or cADP-ribose into the medium for a single granule experiment.
9. To calibrate the intragranular Ca^{2+} changes use ionomycin (20 μM) or A23187 (20 μM) with 1 mM EGTA and 10 mM $CaCl_2$ at the end of experiments.

Results

The zymogen granules preparation

To assess the ability of IP_3 or cADP-ribose to evoke Ca^{2+} release from zymogen granules, it is necessary to isolate a highpurity fraction of granules, but at the same time also evolve a relatively quick isolation protocol so as to preserve the responsiveness of the organelles.

Figure 1 represents a photograph of isolated granules in transmitted light (488 nm). The picture shows mostly single granules, seen as black spheres (zymogen granules effectively absorb light with a wavelength near 500 nm). Figure 2 presents a result of staining of zymogen granules with specific fluorescent dye for acidic organelles LysoTracker. For staining 100 µM of LysoTracker was added to the experimental chamber, and after 5 min the fluorescence image from the same field was recorded (excitation 488 nm, emission 515-560 nm). Most granules stained brightly with this dye, indicating that the intragranular space is acidic. It is possible to use Acridine orange instead of LysoTracker with similar results.

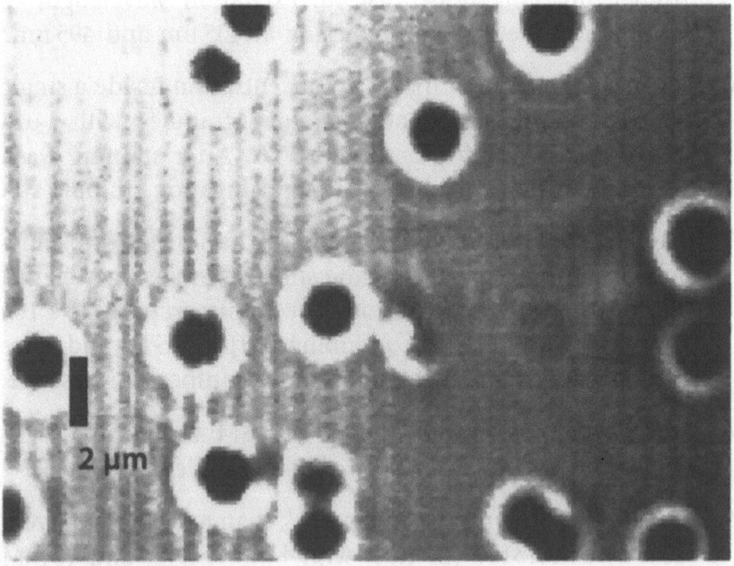

Fig. 1. Transmitted light picture of the specimen of the isolated ZGs from mouse pancreas. Modified from Gerasimenko 1996.

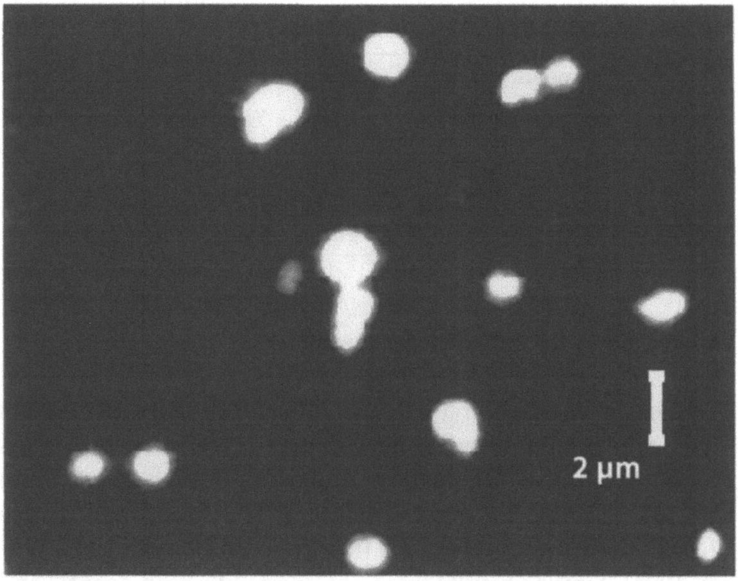

Fig. 2. Lysotracker fluorescence of the ZGs. Modified from Gerasimenko 1996.

The contamination of zymogen granules preparations with endoplasmic reticulum (ER) was tested using staining granules with ER dye 3,3'-dihexyloxacarbocyanine iodide ($DiOC_6(3)$). These granules did not produce any measurable fluorescence, but were brightly stained afterward with LysoTracker. In parallel experiments, DiOC6(3) under the same conditions brightly stained particles in the pancreatic homogenate that was used for isolation of the secretory vesicles (Gerasimenko 1996). In previous experiments on isolated nuclei (Gerasimenko 1995) the bright staining of the nuclear envelope (the outer nuclear membrane has ER properties) with the same concentration of this fluorescent ER marker was shown. Therefore we conclude that such a preparation of the granules is essentially free of ER contamination. This conclusion is reinforced by the inability of thapsigargin, a specific inhibitor of ER Ca^{2+} ATPase, to evoke Ca^{2+} release from the zymogen granules (Gerasimenko 1996).

Activity of α-amylase and total protein concentrations were measured in the granule fraction and the cell homogenate. As shown in Table 1, granule fraction had a 3-fold higher concentration of α-amylase than the homogenate, corresponding to values previously reported (Fuller 1989).

Table 1. α-Amylase and protein content of the homogenate and granular fractions isolated from the mouse pancreas. Modified from Gerasimenko 1996.

Fractions	Volume (ml)	Protein mg/ml	Protein Yield (%)	α-Amylase U/ml	α-Amylase Yield (%)	Enrichment
Homogenate	7	3.9	100	69.4	100	1
ZGs	1	1		208	10.7	3

Ca^{2+} release to the extragranular medium

The measurements of free calcium concentration in the medium outside the granules were done with calcium sensitive indicators Fluo-3 and Indo-1. The extragranular free Ca^{2+} concentration was 160 ± 7 nM. To test the effect of IP_3 the normal d-myo IP_3 or D-myo 3-deoxy-3-fluoro IP_3 were used (Parekh 1993, Wilcox 1994). Similar results were obtained with both agents, and the data were therefore pooled. Application of 5 μM IP_3 (Figure 3) evoked a clear increase in the extragranular Ca^{2+} concentration. IP_3 in a concentration 25 μM evoked an increase of the extragranular Ca^{2+} concentration of 55 ± 5 nM (Gerasimenko 1996). This corresponds approximately to an elevation of about 5 μM of total Ca^{2+} concentration in medium. This represents a decrease of about 0.5 mM in the total Ca^{2+} concentration in the

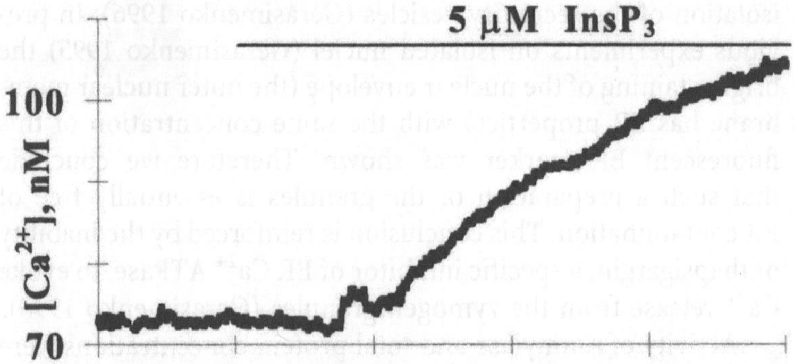

Fig. 3. Effect of IP_3 (5 μM) on isolated ZGs with the extragranular fluorescent probe Fluo-3. Modified from Gerasimenko 1996.

pellet of the granules. Heparin (0.2 mg/ml) abolished the effect of IP_3 (25 μM) in all three experiments carried out.

Application of cADP-ribose from a 1 mM stock to attain a concentration of 25 μM caused a rise in Ca^{2+} concentration of 49 ± 7 nM in the extragranular medium (Figure 4). This corresponds to a 4.3 μM rise of the total Ca^{2+} concentration in the medium, representing a decrease of 0.43 mM in the total Ca^{2+} concentration in the pellet of granules. Both IP_3 and cADP-ribose can be contaminated with Ca^{2+}. Either IP_3 or cADP-ribose (both at 25 μM) added to the same medium caused less than a 5 nM elevation in the free Ca^{2+} concentration. At 10 μM, thapsigargin (Tg), a specific blocker of ER Ca^{2+} ATPases (Thastrup 1990, Clapham 1995, Petersen 1994, Pozzan 1994) did not produce any measurable changes in the fluorescence of Fluo-3 (Figure 5), but IP_3, added afterward induced the usual response.To evaluate the maximal releasable amount of Ca^{2+} from zymogen granules,10 μM of the sample of the granules was put into the camera. Indo-1 (200 μM) and 1 mM EGTA were added. Ca^{2+} was released by addition of 20 μM ionomycin or 0.5 % Triton X-100. The amount of Ca^{2+} released from the granules was estimated by addition of different amounts of Ca^{2+} afterwards and in control experiments. The total releasable Ca^{2+} concentration in the volume of pellet of the granules was 13 ± 3 mM.

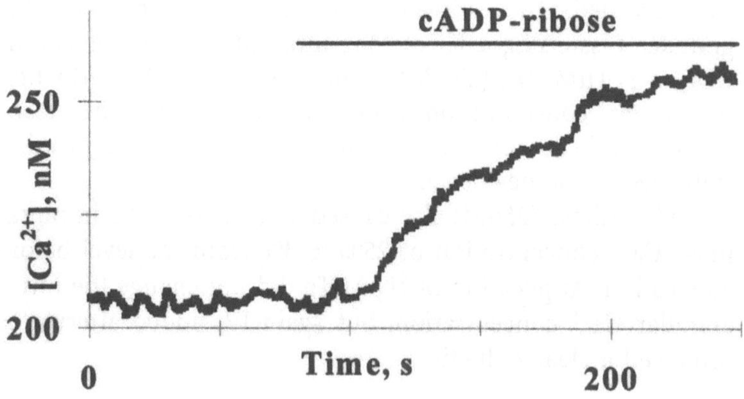

Fig. 4. Effect of cADP-ribose (25 μM) on isolated ZGs with the extragranular fluorescent probe Fluo-3. Modified from Gerasimenko 1996.

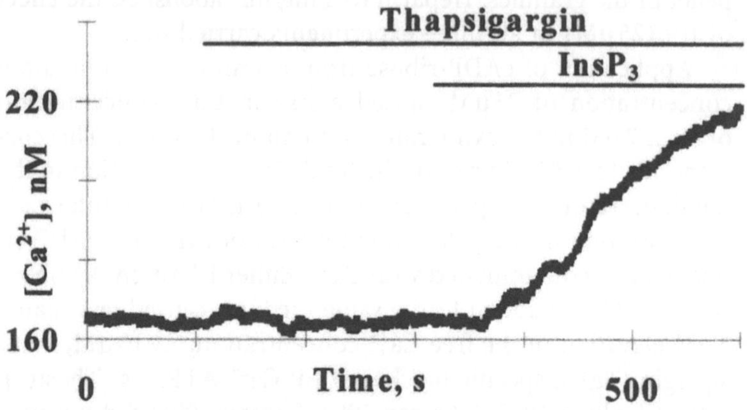

Fig. 5. Lack of thapsigargin 10 μM on isolated ZGs, followed by effect of IP_3 25 μM on isolated ZGs with the extragranular fluorescent probe Fluo-3. Modified from Gerasimenko 1996.

Measurements of changes in the intragranular free Ca^{2+} concentration

To study changes in the free Ca^{2+} concentration inside granules, the granule preparation was incubated with Mag-fura red-acetoxymethyl ester (AM), using the general approach developed by Hofer and Machen (1993). Measurements were done with an exitation wavelength of 488 nm, and emission above 590 nm was recorded. The estimated Ca^{2+} concentration inside the granules (assuming a K_d of Mag-fura red for Ca^{2+} of 17 μM) was 55 ± 4 μM. IP_3 (25 μM) induced a 30 % ± 3% reduction in free Ca^{2+} concentration inside the granules from the prestimation level. Thereafter, the intragranular Ca^{2+} concentration stabilized at the new level.

cADP-ribose (25μM) also caused a decrease in the intragranular Ca^{2+} concentration of 25% ± 3% from the level before application. Application of 10μM Tg did not change the intragranular Ca^{2+} concentration, but again IP_3 added afterwards produced a clear reduction.

Measurements of the free Ca^{2+} concentration inside a single zymogen granules

Single granule measurements were carried out on Mag-fura red-loaded preparations (exitation 488 nm, emission > 590 nm). Fields with granules were chosen using the transparent light option of the confocal microscope, and regions of interest were selected on dark round particles with diameters of about 1 μm. Figure 6 represents examples of such experiments, in which simultaneous fluorescence signals from six isolated granules in the field are presented. The regions chosen coincided with bright particles of the same size as black ones in the transparent light. The estimated free Ca^{2+} concentration inside single

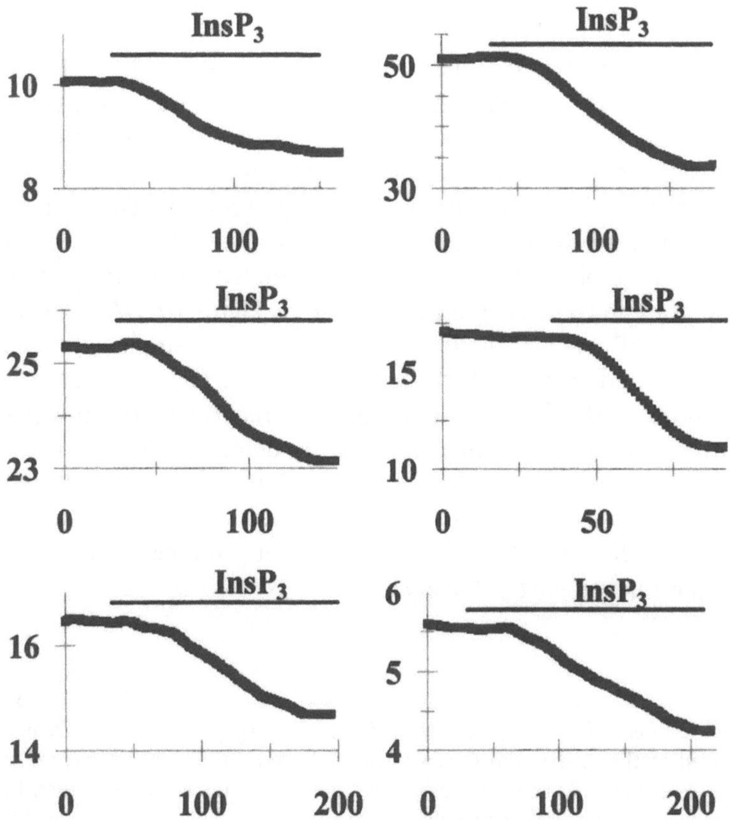

Fig. 6. Effect of IP_3 on six single isolated ZGs, loaded with Mag-fura Red. Modified from Gerasimenko 1996.

granules was $50 \pm 1\,\mu M$. IP_3 was injected into the medium near the granules chosen for measurements and caused a reduction of 25%±3% in the free intragranular Ca^{2+} concentration (Fig. 6). Similar responses were obtained using injection of cADP-ribose (reduction of 30% ± 5%) (Gerasimenko 1996). Addition of $10\,\mu M$ Tg did not evoke any changes (Gerasimenko 1996).

Measurements of the changes of free Ca^{2+} concentration in medium surrounding single granules

Measurements of the release of Ca^{2+} from single granules were caried out using Calcium Green-1 dextran (exitation 488 nm, emission 510-560 nm). Fields with single granules were chosen using the transparent light option of the confocal microscope (Figure 7A), and regions of interest were selected (one close to the secretory granule and another, a control, 10-15 μm away from the granule). Application of the IP_3 was made using the ionophoretic injection (Fig.7B,C). Figure 7D shows recordings of the fluorescence intensity reported by Calcium Green-1 dextran (70 kDa). The upper trace represents the signal measured in the box near the granule (Figures 7A-7C), and the lower trace corresponds to a box that was situated 15 μm away from the granule (outside the field shown in Figures 7A-7C). Injections of IP_3 caused rapid release of Ca^{2+} from single zymogen granule and from small clusters of granules. The transient evaluations of the free Ca^{2+} that could be recorded from the box situated near the granule were hardly detectable a few micrometers away from the granule, and no changes occurred more than 10 μm away. Control injections of IP_3 into the same media but without granules did not produce any changes in the fluorescence intensity.

Fig. 7. Effect of IP_3 on single isolated ZGs, in close vicinity of the granule placed in the media with Calcium Green dextran. A. Schematic picture of the experiment. B. Transmitted light picture. C. Fluorescent light picture. D. Changes of fluorescence of Calcium Green dextran due to injection of IP_3 near the granule (upper trace) and lack of changes in control box 10 μm away from the granule (lower trace). Reproduced with permission from Cell Press (Gerasimenko 1996).

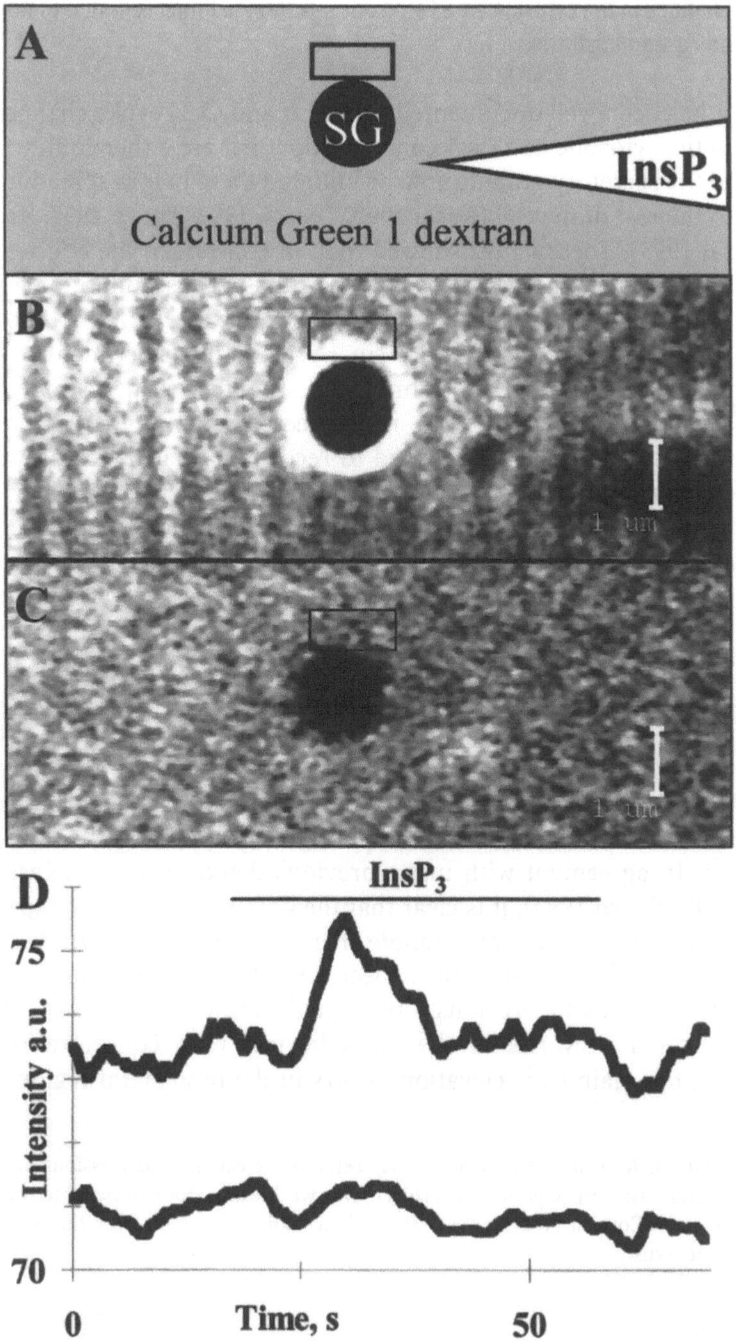
A
SG
$InsP_3$
Calcium Green 1 dextran
B
1 um
C
1 um
D
$InsP_3$
75
Intensity a.u.
70
0
Time, s
50

Changes in the cytosolic free Ca^{2+} concentration in intact cells due to ACh or Tg administration

It has been well documented that ACh and CCK evoke changes in the cytosolic free Ca^{2+} concentration that are either confined to the secretory granule area or initiated there before spreading to the rest of the cell (Kasai 1990, Toescu 1992, Thorn 1993, Kasai 1993). Toescu 1992 showed that, in contrast to the effect of the secretagogues, the specific ER Ca^{2+} ATPase inhibitor Tg (Thastrup 1990) primarily evoked Ca^{2+} release from the basolateral regions of pancreatic acinar cells. In the experiments reported by Toescu (1992), a brief ACh stimulus was first applied, causing a distinct Ca^{2+} rise in the secretory granule area, and thereafter the cell was allowed to recover so that the cytosolic free Ca^{2+} concentration in the secretory granule area was back to normal before application of Tg. Nevertheless, it could be argued that the failure of Tg in these experiments to cause Ca^{2+} release in the secretory granule area (Toescu 1992) might have been due to the previous ACh-evoked Ca^{2+} release in that region.

In view of the data shown above indicating that Tg cannot release Ca^{2+} from isolated zymogen granules, it was of interest to test in the intact cells, using the confocal microscope, where the primary Tg-evoked Ca^{2+} release occurs. Gerasimenko 1996 shows the effect of a brief ionophoretically applied ACh stimulus. In agreement with many previous demonstrations (Kasai 1993, Thorn 1993), it is clear that the Ca^{2+} rise occurs very specifically in the secretory granule area. The effect of Tg (10 μM) in a cell that had not previously been stimulated with agonist is in agreement with previously published data (Toescu 1994). Tg evokes a slow rise in the cytosolic free Ca^{2+} concentration and the main Ca^{2+} elevation occurs in the basolateral regions.

Acknowledgements. We thank Dr. A. Tepikin for extensive discussions and helpful advice throughout this project. This work was supported by a Medical Research Council Programme Grant and an Equipment Grant from the Wellcome Trust.

References

Berridge MJ (1993) Inositol trisphosphate and calcium signalling. Nature 361:315-325

Blondel O, Bell GI, Seino S (1995) Inositol 1,4,5-trisphosphate receptors, secretory granules and secretion in endocrine and neuroendocrine cells. Trends Neurosci. 18:157-161

Cancela JM, Churchill GC, Galione A (1999) Coordination of agonist-induced Ca^{2+}-signalling patterns by NAADP in pancreatic acinar cells. Nature 398:74-76

Clapham DE (1995) Calcium signalling. Cell 80:259-268

Clemente F, Meldolesi J (1975) Calcium and pancreatic secretion. I. Subcellular distribution of calcium and magnesium in the exocrine pancreas of guinea pig. J. Cell Biol. 65:88-102

Fasolato C, Zottini M, Clementi E, Zacchetti D, Meldolesi J, Pozzan T (1991) Intracellular Ca^{2+} pools in PC12 cells: three intracellular pools are distinguished by their tumover and mechanisms of Ca^{2+} accumulation, storage and release. J. Biol. Chem. 266:20159-20167

Fuller CM, Deetjen HH, Piiper A, Schulz I (1986) Secretagogue and second messenger-activated Cl^- permeabilities in isolated pancreatic zymogen granules. Pflugers Arch. 415:29-36

Gerasimenko OV, Gerasimenko JV, Tepikin AV, Peteresen OH (1995) ATP-dependent accumulation and inositol trisphosphate- or cyclic ADP-ribose-mediated release of Ca2+ from the nuclear envelope. Cell 80:439-444

Gerasimenko OV, Gerasimenko JV, Belan PV, Petersen OH (1996) Inositol trisphosphate and cyclic ADP-ribose-mediated release of Ca^{2+} from single isolated pancreatic zymogen granules. Cell 84:473-480

Gorelick FS, Jamieson JD (1994) The pancreatic acinar cell: structure-function relationships. In Physiology of the Gastrointestinal Tract, Third edition, L.R.Johnson,ed. (New York:Raven Press) :1353-1376

Hofer AM, Machen TE (1993) Technique for in situ measurement of calcium in intracellular inositol 1,4,5-trisphosphate-sensitive stores using the fluorescent indicator mag-fura-2. Proc.Natl. Acad. Sci. USA 90:2598-2602

Kasai H (1995) Pancreatic calcium waves and secretion. Ciba Foundation Symp. 188:104-120

Kasai H, Augustine GJ (1990) Cytosolic Ca2+ gradients trigering unidirectional fluid secretion from exocrine pancreas. Nature 348:735-738

Kasai H, Petersen OH (1994) Spatial dynamics of second messengers: IP3 and cAMP as long-range and associative messengers. Trends Neurosci. 17:95-101

Kasai H, Li YX, Miyashita Y (1993) Subcellular distribution of Ca^{2+} release chanels underlying Ca^{2+} waves and oscilations in exocrine pancreas. Cell 74:669-677

Kern HF (1993) Fine structure of the human exocrine pancreas. In The Pancreas: Biology, Pathobiology and Disease, Second Edition, V.L.W. Go, ed. (New York: Raven Press):9-19

Maryama Y, Petersen OH (1994) Delay in granular fusion evoked by repetitive cytosolic Ca^{2+} spikes in mouse pancreatic acinar cells. Cell Calcium 16:419-430

Maryama Y, Inooka G, Li YX, Miyashita Y, Kasai H (1993) Agonist-induced localized Ca2+ spikes directly triggering exocytotic secretion in exocrine pancreas. EMBO J. 12:3017-3022

Mogami H, Tepikin AV, Petersen OH (1998) Termination of cytosolic Ca^{2+} signals: Ca^{2+} reuptake into intracellular stores is regulated by the free Ca^{2+} concentration in the store lumen. EMBO J. 17: 435-442

Nakagaki I, Sasaki S, Shiguma M, Imai Y (1984) Distribution of elements in pancreatic exocrine cells determined by electron probe X-ray analysis. Pfluger Arch. 401:340-345

Nathanson MH, Fallon MB, Padfield PJ, Maranto AR (1994) Localisation of the type 3 inositol 1,4,5-trisphosphate receptor in the Ca2+ wave trigger zone of pancreatic acinar cells. J. Biol. Chem. 269:4693-4696

Nicaise G, Maggio K, Thirion S, Horoyan M, Keicher E (1992) The calcium loading of secretory granules: a possible key event in stimulus-secretion coupling. J. Biol. Cell. 75:89-99

Nguyen T, Chin WC, Verdugo P (1998) Role of Ca2+/K+ ion exchange in intracellular storage and release of Ca2+. Nature, 395:908-912

Parekh AB, Fogut M, Lubbert H, Stuhmer W (1993) Ca2+ oscillations and Ca2+ influx in Xenopus oocytes expressing a novel 5-hydroxytryptamine receptor. J. Physiol. (Lond.) 469:653-671

Parekh A, Penner R (1997) Store depletion and calcium influx. Physiol. Rev. 77:901-930

Petersen OH (1980) In: The electrophysiology of gland cells. Academic Press, London, New York, pp

Petersen OH (1992) Stimulus-secretion coupling: cytoplasmic and calcium signals and the control of ion channels in exocrine acinar cells. J. Physiol. 448: 1-51

Petersen OH, Petersen CCH, Kasai H (1994) Calcium and hormone action. Annu.Rev. Physiol. 56:297-319

Pozzan T, Rizzuto R, Volpe P, Meldolesi J (1994) Molecular and cellular physiology of intracellular calcium stores. Physiol. Rev. 74:595-636

Saurin AJ, Hamlett J, Clague MJ, Pennington SR (1996) Inhibition of mitogen-induced DNA synthesis by bafilomycin A_1 in Swiss 3T3 fibroblasts. Biochem. J. 313:65-70

Smith C, Neher E (1997) Multiple forms of endocytosis in bovine adrenal chromaffin cells. J. Cell Biol. 139:885-894

Steinman RM, Mellman IS, Muller WA, Cohn ZA (1993) Endocytosis and the recycling of plasma membrane. J. Cell Biol. 96:1-27

Tepikin AV, Voronina SG, Gallacher DV, Petersen OH (1992a) Acetylcholine-evoked increase in the cytoplasmic Ca^{2+} concentration and Ca^{2+} extrusion measured simultaneously in single mouse pancreatic acinar cells. J. Biol. Chem. 267:3569-3572

Tepikin AV, Voronina SG, Gallacher DV, Petersen OH (1992b) Pulsatile Ca^{2+} extrusion from single pancreatic acinar cells during receptor-activated cytosolic Ca^{2+} spiking. J. Biol. Chem. 267:14073-14076

Thastrup O, Cullen PJ, Dvorak BK, Hanley MR, Dawson AP (1990) Thapsigargin, a tumor promoter, discharges intracellular Ca2+ stores by specific inhibition of the endoplasmic reticulum Ca2+ ATPase. Proc. Natl. Acad. Sci. USA 87:2466-2470

Thorn P, Lawrie AM, Smith PM, Gallacher DV, Petersen OH (1993) Local and global cytosolic Ca2+ oscillations in exocrine cells evoked by agonists and inositol trisphosphate. Cell 74:661-668

Toescu EC, Lawrie AM, Petersen OH, Gallacher DV (1992) Spatial and temporal distribution of agonist-evoked cytoplasmic Ca2+ signals in exocrine acinar cells analysed by digital image microscopy. EMBO J. 11:1623-1629

Toescu EC, Petersen OH (1994) The thapsigargin-evoked increase in [Ca2+]i involves an IP3 - dependent Ca2+ release process in pancreatic acinar cells. Pflugers Arch. 427:325-331

Wilcox RA, Challiss RA, Traynor JR, Fauq AH, Ognayanov VI, Kozikovski AP, Nahorski SR (1994) Molecular recognition at the myo-inositol 1,4,5-trisphosphate receptor: 3-position substituted myo-inositol trisphosphate analogues reveal the binding and Ca2+ release requirements for high affinity interaction with the myo-inositol 1,4,5 receptor. J. Biol. Chem. 269:26815-26821

Yoo SH, Albanesi JP (1990) Inositol 1,4,5-trisphosphate-triggered Ca2+ release from bovine adrenal medullary secretory vesicles. J. Biol. Chem. 265:13446-13448

Zen K, Biwersi J, Periasamy N, Verkman AS (1992) Second messengers regulate endosomal acidification in Swiss 3T3 fibroblasts. J. Cell Biol. 119:99-110

Ca^{2+} Measurement in Endosomes of Intact Cells

JULIA V. GERASIMENKO

Introduction

The cellular calcium content depends on a balance between uptake and release. It has been generally accepted that Ca^{2+} is released from cells via plasma membrane calcium pumps and in some cases Na^+-Ca^{2+} exchangers whereas uptake is mediated by a variety of plasma membrane calcium channels (Petersen 1994, Pozzan 1994; Parekh and Penner 1997). The calcium transport into cells by endocytosis has been overlooked and amounts of calcium that could be transported in this way has been underestimated. Little attention has been paid so far to the possibility that substantial amounts of calcium could be exported and imported via exo- and endocytosis, although it has been recently shown that exocrine cells can release significant quantities of calcium during agonist-induced exocytosis (Belan 1998).

Endocytosis is the general term used for the internalization of extracellular substances by invaginations of the plasma membrane and subsequent pinching off to create vesicles called endosomes. This process occurs in all cells (Mellman 1996). Fibroblasts and macrophages internalize more than 200% of their entire surface area every hour (Steinman 1993). There is also evidence for several fast exocytosis-coupled forms of endocytosis (Smith 1997).

The lumen of the endosomes is known to be acidic mainly due to operation of a vacuolar-type proton ATPase (Mellman 1986). Acidification can be specifically blocked by bafylomycin

Julia V. Gerasimenko, The University of Liverpool, The Physiological Laboratory, Crown Street, P.O. Box 147, Liverpool, L69 3BX, UK
(*phone* +44-151-794-5357; *fax* +44-151-794-5327;
e-mail jvg1@Liverpool.ac.uk)

(Bowman 1988). The pH of "early" endosomes is in the range of 6-6.5, then decreases progressively to 5.5-6 for "sorting" endosomes, and "late" endosomes and lysosomes have the pH 4.5-5.5 (Zen 1992).

The proton ATPase is electrogenic (Mellman 1986) and proton accumulation inside the endosomes can therefore only occur if the proton movement is accompanied by anion movement or is associated with exchange of another cation. It has been suggested that the H^+ accumulation is due in part to Cl^- inward movement and in part to outward Na^+ or K^+ movement (Mellman 1986; Zen 1992).

The changes of fluorescence in the endosomes have been monitored using confocal fluorescent microscopy. Calcium or pH calibrations have been used to perform free Ca^{2+} or pH measurements inside endosomes. Free Ca^{2+} measurements performed for the first time in endosomes, allow to investigate calcium changes in endosomes after different time of incubation with fluorescent dyes and to investigate the fate of calcium taken up by endocytosis.

The early endosomes have already lost most of the Ca^{2+} they had taken up from the external solution confirming the suggested high Ca^{2+} permeability of the endosomal membrane (Diaz 1989). It was confirmed that bafylomycin prevents acidification, but it was also found that bafylomycin completely blocked Ca^{2+} loss from the endosomes. Not only calcium release is dependent on acidification but acidification is also dependent on external calcium. Reduction in external calcium inhibits acidification. Endosomal acidification and loss of Ca^{2+} from the endosome occur within the same time frame. The fact that the endosomal Ca^{2+} loss depends on acidification and vice versa suggests that Ca^{2+} release from endosomes is an important element in the endosomal acidification process. The amount of calcium imported into cells by endocytosis is considerable and is in the same range as calcium influx into cells through calcium channels.

Outline

The aim of this study is to characterize Ca^{2+} loss from endosomes and to investigate its possible relationship to endosomal acidification. Ca^{2+} sensitive or H^+ -sensitive fluorescent indicators were placed in the extracellular solution and were subsequently trapped inside fibroblasts by endocytosis.

Materials

- CO_2 incubator 37°C
- 200 µl perfusion chamber connected with syringes
- Tissue culture petri dishes, 35 X 10 mm
- 22 mm round glass coverslips
- Laser scanning confocal microscope (Noran Instruments, USA) with X 60 oil objective (NA 1.4, Nicon)
- Swiss 3T3 Albino fibroblasts (European Collection)
- Dulbecco's modification of Eagle's medium (DMEM) supplemented with 10 % fetal bovine serum, 100 µU/ml penicillin and 100 µg/ml streptomycin (Gibco BRL)
- Ionomycin, nigericin, bafilomycin, EGTA (Sigma)
- Fluoresceine isothiocyanate (FITC)-dextran 9.3 kDa, rhodamine B, rhodamine B isothiocyanate dextran 9.3 kDa (Sigma), Oregon Green 488 BAPTA-5N (Molecular Probes)

Bath solution

- 40 mM NaCl
- 4.7 mM KCl
- 1.1 mM $MgCl_2$
- 50 mM Hepes
- 10 mM glucose
- 2 mM Ca Cl_2
- adjust to pH 7.4

Procedure

Cell culture and endocytic labelling

1. Plate Swiss 3T3 Albino fibroblasts on 22 mm round glass coverslips in DMEM with 2 mM glutamine and 10% FBS and incubate at 37 °C with 10% CO_2 and use 2-4 days after plating.
2. For fluorescent labelling of the endocytic vesicles incubate the cells for 5, 10 or 20 minutes in a DMEM containing 100 μM of the fluorescent dye (fluorescein isothiocyanate-dextran 9.3 kD, rhodamine B, rhodamine B isothiocyanate-dextran 9.3 kD or Oregon Green 488 BAPTA-5N, in a CO_2 incubator at 37 °C.
3. After incubation transfer the coverslips to a 200 μl perfusion chamber where coverslips are placed in contact with the objective.
4. Perfuse the cells continuously with a bath solution at 37 °C or at room temperature.

Measurements of the pH in endosomes

For calculation of the endosomal pH values use 2-point calibration procedure: the cells with FITC-dextran labelled endosomes were exposed to a high K buffer (100 mM Na^+ in a bath solution replaced by K^+) containing 5 μM nigericin to equilibrate endosomal and extracellular pH to 7.4 and then to 5.3 (Zen 1992) (Fig.2). Use the excitation wavelength for fluorescein isothiocyanate-dextran 488 nm and the emission wavelength 530 nm.

Measurements of free calcium concentration in endosomes

For calculations of calcium concentration use calibrations of the intensity of calcium indicator inside the endosomes with ionomycin (20 μM) in the presence of nigericin (5 μM) and 2 mM $CaCl_2$ and then 2 mM EGTA in a physiological buffer (bath solution) (Fig 5.). The formula for calculation of calcium concentration is:

$$[Ca^{2+}] = K_d * (F - F_{min})/(F_{max} - F)$$

- K_d= 20 μM, for Oregon Green BAPTA 5N (Molecular Probes).
- F - intensity of fluorescense of the endosomes at the beginning of the calibration experiment;

- F_{max} - intensity of fluorescense of the endosomes after addition of ionomycin (20 μM), nigericin (5 μM) and 2 mM $CaCl_2$ to the bath solution;
- F_{min} - intensity of fluorescense of the endosomes after addition of 2 mM EGTAto the bath solution.

Confocal microscopy
The fluorescence measurements can be performed with the help of Noran Odyssey laser scanning confocal microscope (Noran Instruments, USA). Use the high aperture PlanApo x60, NA 1.4 (Nikon) objective to get enough fluorescence signal. For excitation use 488 nm argon laser wavelength and emission wavelengths for Oregon Green BAPTA 5N and FITC are 530 nm and 600 nm for rhodamine B. Use largest slit 100 μm in slow mode (sample time 3200 nsec) to allow use of integration 8 (frame speed 0.5 per sec).

Note: These settings are important to resolve very low fluorescent signals without bleaching and possible light damage to endosomes.

Results

Time courses of endosomal acidification

The highly fluorescent endocytic vesicles are present in the peripheral part of the cell close to the plasma membrane after incubation with FITC-dextran for 5-20 minutes at 37 °C (Fig. 1 ab modified from Gerasimenko 1998). No fluorescence or endosomes could be seen in the nuclear area.

The pK_a of FITC-dextran is about 6.4 with maximal sensitivity in the pH range 6.0 -7.2 (Molecular Probes). Using the pH calibrations with the protonophore nigericin it has been shown that after 3 min incubation of cells with FITC-dextran the pH in the endosomes decreased from the value of 7.4 in the external solution to 6.9 ±0.2. After 5 minutes of incubation the pH inside the endocytic vesicles was 6.6± 0.2, after 10 min 6.0 ± 0.2 and after 20 min incubation the pH decreased to 5.7 ± 0.2 (Fig. 2 modified from Gerasimenko 1998).

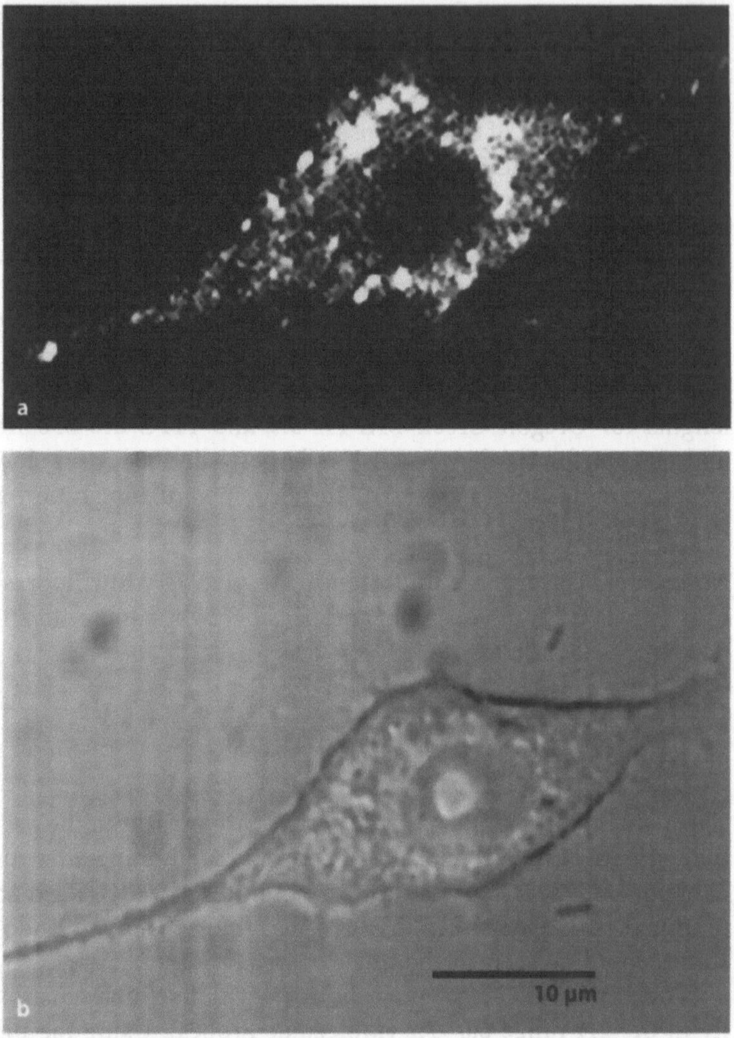

Fig. 1. Distribution of the pH-sensitive probe FITC-dextran. **a** The localisation of the FITC-dextran in the peripheral cytoplasm of a Swiss 3T3 fibroblast. The confocal image was taken folowing incubation for 10 min at 37 °C. **b** Transmitted light picture of the same cell. Bar represents 10 μM.

Time course of calcium release from endosomes

To measure calcium in endosomes the cells were incubated in a solution containing the Ca^{2+} -sensitive dye Oregon Green 488 BAPTA-5N which is pH insensitive at pH higher than 6.0 (Molecular Probes) and decreases brightness by 10% when pH changes from 6.0 to 5.0. There is a peripheral distribution of

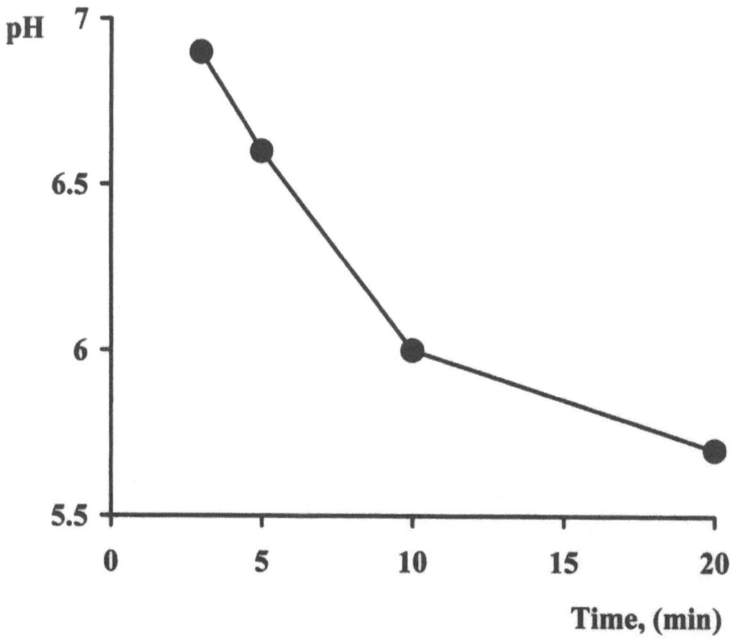

Fig. 2. The dependence of endosomal pH on the time of incubation with FITC-dextran.

round shaped structures with a size less than 1 μm or clusters of such structures but no fluorescence or endosomal structures are seen in the nuclear region after incubation (Fig 3ab modified from Gerasimenko 1998).

The dissociation constant (K_d) of the Oregon Green BAPTA 5N for Ca^{2+} is 20 μM (Molecular Probes). The level of free Ca^{2+} concentration in the endosomes has been obtained by calibration with addition of the Ca^{2+} ionophore ionomycin in the presence of nigericin first in a buffer with 2 mM $CaCl_2$ and then in a buffer with 2 mM of the Ca^{2+} chelator EGTA. Fig. 4 (modified from Gerasimenko 1998) presents the time course of changes in the endosomal Ca^{2+} concentration. After 3 min of incubating cells in a Na Hepes buffer containing 2 mM $CaCl_2$ the free Ca^{2+} concentration in the endosomes was 28.6 μM ± 2. After 5 min of incubating cells in a solution containing 2 mM $CaCl_2$ the free Ca^{2+} concentration in the endosomes changed to 8.5 μM ± 1.8. After 10 min it had decreased to 3.9 μM ± 1.2 and after 20 min the value was 3.0 μM ± 0.8. These data indicate that ex-

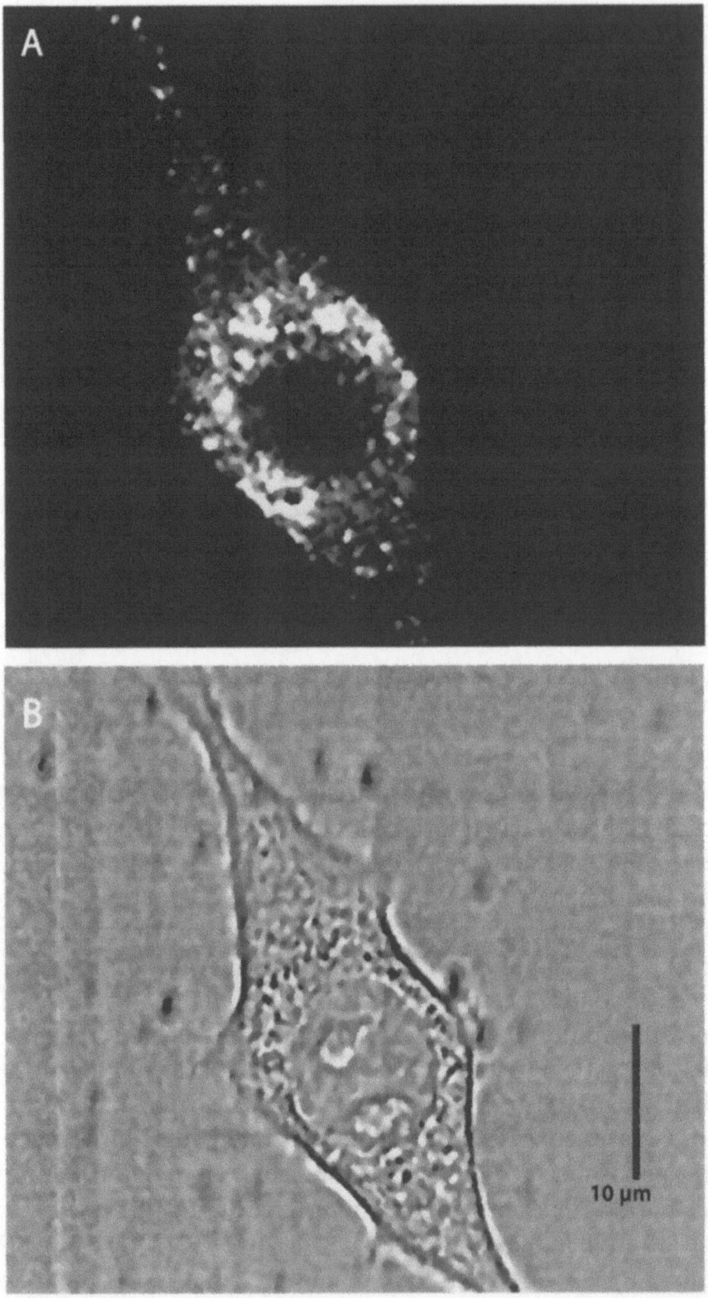

Fig. 3. Ca^{2+} concentration in endosomes measured by the calcium sensitive fluorescent dye Oregon Green 488 BAPTA-5N. **A** Confocal image of the fluorescence of Oregon Green 488 BAPTA-5N in fibroblast after incubation for 10 min at 37 °C. **B** Transmitted light picture of the same cell. Bar represents 10 µM.

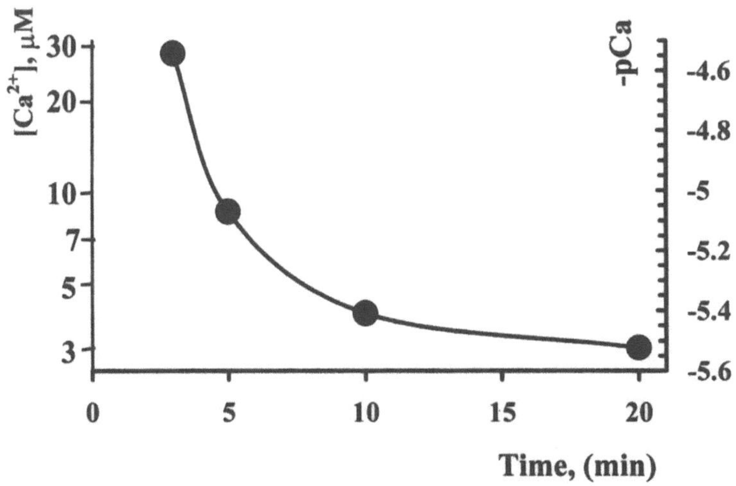

Fig. 4. The dependence of the free Ca^{2+} concentration in the endosomes on the incubation time with Oregon Green 488 BAPTA-5N.

ternal calcium taken up by endocytosis has already been lost to the cytosol within the first 3 minutes.

Ratiometric measurements of free calcium in endosomes

The endosomal Ca^{2+} concentration was also measured using the ratio of the fluorescence intensities of rhodamine B and Oregon Green BAPTA 5N. Fig. 5 shows the typical changes of this ratio during calibration with Ca and EGTA containing solutions in the presence of ionomycin and nigericin. Using these ratio measurements it has been calculated that the free Ca^{2+} concentration in the endosomes after 20 min incubation was 2.9 µM ± 1.2 i.e. in the same range as measured using Oregon Green 488 BAPTA-5N fluorescence only. These data confirm that most of the free Ca^{2+} taken up by endocytosis has already been lost from the endosomes to the cytosol within the first few minutes.

The relationship between endosomal acidification and Ca^{2+} loss

It is well established that bafylomycin, the specific inhibitor of the vacuolar type proton ATPase (Bowman 1988; Bode 1996) blocks endosomal acidification (Clague 1994; Saurin 1996). The maximal inhibitory effect of bafylomycin on vacuolar ATPases appears at 100 nM but even at a concentration of

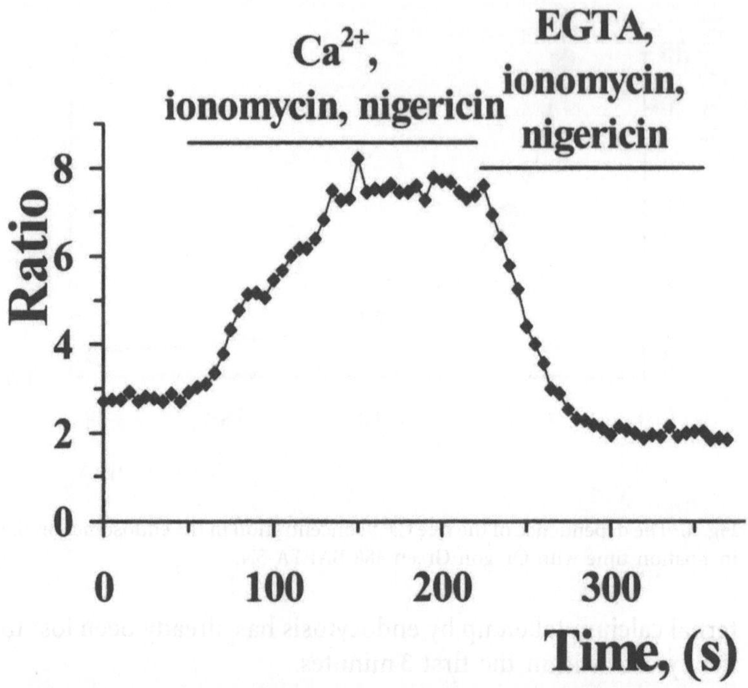

Fig. 5. Measurement of the free endosomal Ca^{2+} concentration by the ratio of the fluorescences of Oregon Green 5N and rhodamine B isothiocyanate dextran in Swiss 3T3 fibroblasts. Cells were incubated with this combination of indicators for 20 minutes at 37 °C. Calibration of the ratio of the fluorescences was performed using applications of ionomycin (20 μM) and nigericin (5 μM) in a buffer with 2 mM $CaCl_2$ and then in a buffer with 2 mM EGTA.

1 μM it has no effect on other ATPases (Bowman 1988). We therefore compared the endosomal Ca^{2+} concentration in cells incubated with 100 nM bafylomycin with that in control cells (Gerasimenko 1998). The introduction of a high Ca^{2+} concentration in the presence of ionomycin normally evoked a marked increase in the fluorescence intensity of the endocytosed Oregon Green 5N which could then be markedly reduced by the calcium chelator EGTA. In contrast Ca^{2+} and ionomycin failed to evoke any increase in the fluorescence from the endocytosed marker in the cells that had been incubated with bafylomycin demonstrating that the indicator was saturated with Ca^{2+}. Subsequent introduction of EGTA reduced the fluorescence. The free Ca^{2+} concentration in the endosomes in the bafylomy-

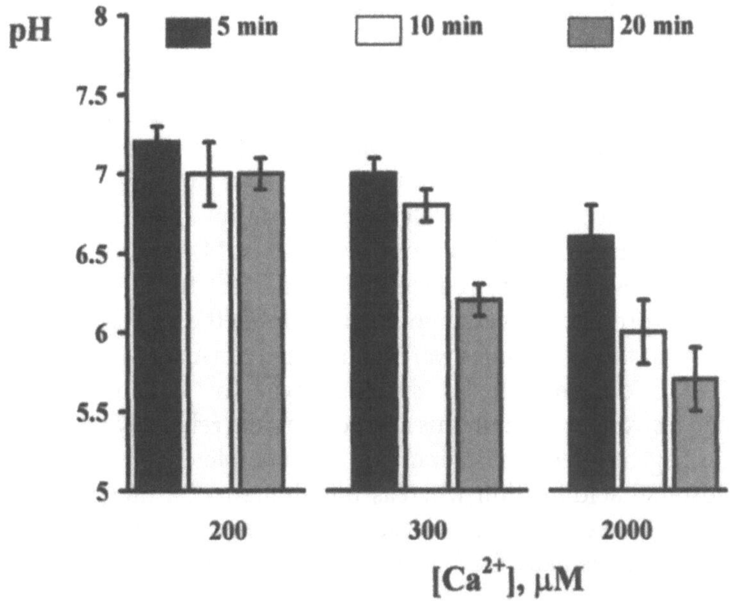

Fig. 6. The effect of changing the external Ca^{2+} concentration on the endosomal pH in Swiss 3T3 fibroblasts.

cin-treated fibroblasts was therefore high (> 200 µM). These experiments indicate that blockade of the proton ATPase prevents loss of Ca^{2+} from endosomes (Gerasimenko 1998).

Dependence of acidification on external calcium concentration

The experiments described in Fig. 2 were carried out in the presence of 2 mM Ca^{2+} in the external solution. The same type of experiment has been done for the cells incubated in solutions with Ca^{2+} concentrations of 200 µM, 300 µM or 2 mM. As seen in Fig. 6 the degree of endosomal acidification at 5,10 and 20 minutes was similar to the control results shown in Fig. 2 when the external Ca^{2+} concentration was 2 mM. When the external Ca^{2+} concentration was reduced to 300 µM the degree of acidification was considerably inhibited and at an external Ca^{2+} concentration of 200 µM there was no acidification at all. These results indicate that acidification can only occur when there is an initial high Ca^{2+} concentration in the endosomes.

Inhibition of calcium release from endosomes by calcium channel blockers

Although we have no information about the nature of the Ca^{2+} release pathway in the endosomal membrane, we nevertheless attempted to interfere with the Ca^{2+} release by employing Ca^{2+} channel blockers. Relatively high concentrations of the L-type Ca^{2+} channel blockers verapamil (250 μM) and nifedipine (50 μM) were able to reduce the loss of Ca^{2+} from endosomes. In the presence of verapamil or nifedipine the Ca^{2+} concentrations in the endosomes were higher than normal (Gerasimenko 1998). Verapamil, the less specific L-type Ca^{2+} channel blocker, had a stronger effect in this respect than the relatively more specific agent nifedipine. Verapamil also completely prevented endosomal acidification whereas nifedipine had a much smaller effect. These data again suggest that there is a correlation between endosomal Ca^{2+} loss and acidification (Gerasimenko 1998).

The dependence of the rate of endocytosis on the external Ca^{2+} concentration

The rate of endocytosis can be calculated by measuring the uptake of rhodamine B isothiocyanate dextran into the fibroblasts (Fig. 7 modified from Gerasimenko 1998). The amount of endocytosed rhodamine-dextran increased linearly with the time. The rate of uptake was about 0.1 μM per minute (at external concentration of rhodamine-dextran of 100 μM). At external calcium concentration of 2 mM, that would correspond to the rate of 2 μM per minute of unstimulated calcium uptake.

These experiments were then carried out at different external Ca^{2+} concentrations. The rate of endocytosis was not very sensitive to reductions of the external Ca^{2+} level. Thus a reduction of the external Ca^{2+} concentration from 1 mM to 100 μM only resulted in a marginal attenuation of the endocytosis rate. However, when the external Ca^{2+} concentration was dramatically reduced by adding the Ca^{2+} chelator EGTA to a nominally Ca^{2+} -free solution, then the rate of endocytosis fell sharply (Gerasimenko 1998).

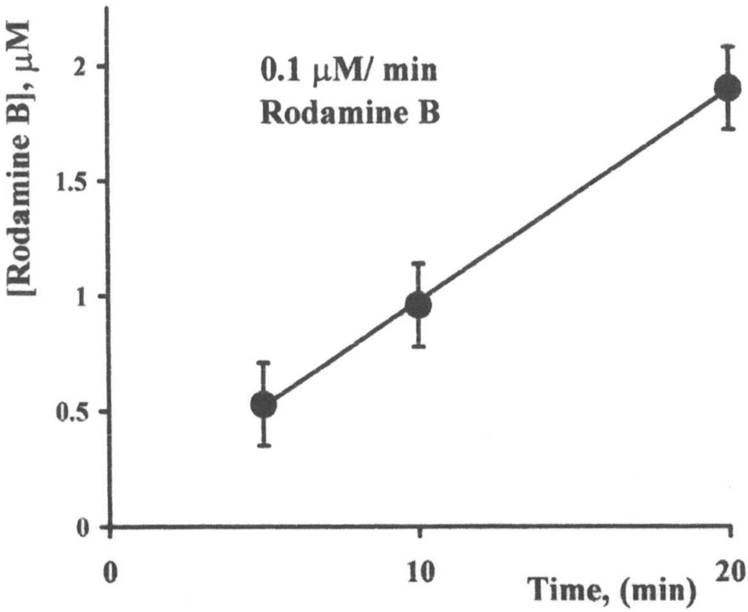

Fig. 7. The rate of endocytosis of rhodamine-dextran in Swiss 3T3 fibroblasts.

Flash photolysis of caged calcium in endosomes

The availability of caged compounds allows to extend the possibilities of studies of free calcium in endosomes. UV application can very quickly, in a few milliseconds change affinity of NP-EGTA to calcium, resulting in immediate elevation of free calcium in endosomes. One of the possible experiments shown in Fig. 8. Fibroblasts were preincubated with calcium-sensitive dye Oregon Green Bapta 5N and NP-EGTA. After that UV light of the UV laser of Noran confocal was applied for a few seconds and excitation returned to 488 nm. Free calcium was quickly released during UV flash - sharp increase at 24 second. This experiment allows the direct monitoring of calcium release from the endosomes - exponential decrease of the free calcium after the sharp increase (Gerasimenko 1998). These data confirm previous conclusions about fast loss of calcium from endosomes, possibly through some kind of calcium channels.

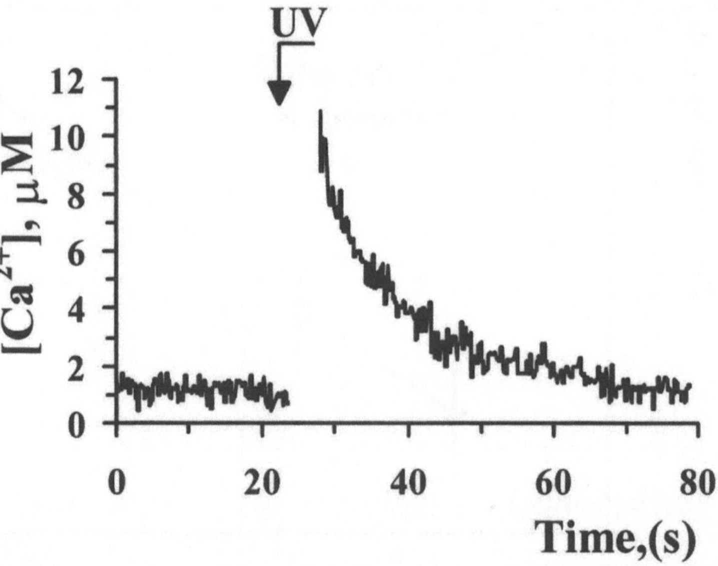

Fig. 8. Flash photolisis of endocytosed NP-EGTA. The endosomes of Swiss 3T3 fibroblasts were loaded with NP-EGTA and Oregon Green 488 BAPTA-5N. Arrow represents onset of UV flash triggering release of Ca^{2+} within the endosomes.

Conclusions

1. extracellular Ca^{2+} taken up by endocytosis is quickly lost from the endosomes to the cytosol,
2. the time course of this Ca^{2+} loss is similar to that of endosomal acidification,
3. when the amount of Ca^{2+} taken up by endocytosis is reduced by reducing the external Ca^{2+} concentration the endosomes become less acidic and
4. when the vacuolar proton ATPase is inhibited, the normally occurring Ca^{2+} loss from the endosomes is blocked.

Comments

It has been generally accepted that the vacuolar proton ATPase is electrogenic and that proton accumulation in endosomes therefore requires either Cl^- uptake and/or K^+ release (Mellman 1986). However, most cells have a considerably lower Cl^- concentration intracellularly than extracellularly and the K^+ concentration in the extracellular fluid that is taken up by endocytosis is much lower than in the cytosol (Petersen, 1980). In the absence of information about the electrical potential difference across the endosomal membrane, it is not clear whether proton accumulation can be satisfactorily explained by accompanying Cl^- transport or K^+ exchange. Our data indicate that proton uptake may, to a large extent, be dependent on Ca^{2+} release from the endosomes. When the extracellular Ca^{2+} concentration was lowered to 200 µM there was essentially no endosomal acidification (Fig. 6). Since Ca^{2+} can control opening of K^+ and Cl^- channels (Petersen 1992), these experiments could also be interpreted by assuming a more indirect role for Ca^{2+} in, for example, controlling K^+ and/or Cl^- permeability. However, taking into account our experimental finding that there is no Ca^{2+} loss from the endosomes when the vacuolar proton ATPase has been blocked by bafylomycin, the simplest interpretation seems to be that proton accumulation is dependent on Ca^{2+} release. During the period of measurements (3-20 min after the beginning of endocytosis), it is apparent that the Ca^{2+} concentration is reduced by a factor of about 10, whereas the H^+ concentration is increased by a factor of 15. The exact buffering situation in the endosomes is unknown at this stage and can therefore not assess the H^+/Ca^{2+} flux ratio. These data strongly suggest that H^+ accumulation is dependent on Ca^{2+} release, although additional charge compensation by monovalent ions, possibly through Ca^{2+} -dependent channels cannot be excluded. Use of caged calcium together with simultaneous measurements of free calcium and pH in endosomes could clarify the relationship between calcium loss and acidification (Gerasimenko 1998).

These data indicate that endocytosis can result in a considerable import of Ca^{2+} into the cell and that this Ca^{2+} is released into the cytosol. If the endoplasmic reticulum (ER) has been depleted of Ca^{2+}, for example by stimulation with an inositol trisphosphate-generating agonist (Berridge 1993), then Ca^{2+} released from endosomes could be effectively taken up into the endoplasmic reticulum by the thapsigargin-sensitive Ca^{2+} ATPase pump (Mogami 1998). We have recently shown that a considerable amount of Ca^{2+} can be exported from cells via agonist-evoked exocytosis (Belan 1998). In addition to Ca^{2+} extrusion by plasma membrane Ca^{2+} pumps (Tepikin 1992a,b) and Ca^{2+} influx via various Ca^{2+} entry channels (Parekh and Penner 1997) we must now also include Ca^{2+} uptake and extrusion via endo and exocytosis into our overall picture of cellular calcium homeostasis.

Acknowledgements. This work was supported by an MRC Programme Grant (G 8801575).

References

Belan PV, Gardner J, Gerasimenko O, Gerasimenko J, Lloyd-Mills C, Petersen OH, Tepikin AV (1998) Isoproterenol evokes extracellular Ca^{2+} spikes due to secretory events in salivary gland cells. J. Biol. Chem. 273: 4106-4111

Berridge MJ (1993) Inositol trisphosphate and calcium signalling. Nature 361:315-325

Bode H-P, Himmen A and Göke B (1996) Evidence for vacuolar-type proton pumps in nonmitochondrial and inositol 1,4,5-trisphosphate-sensitive calcium stores of insulin-secreting cells. Pflügers Arch. 432:97-104

Bowman EJ, Siebers A, Altendorf K (1988) Bafilomycins: a class of inhibitors of membrane ATPases from microorganisms, animal cells and plant cells. Proc. Natl. Acad. Sci. U.S. 85: 7972-7976

Clague MJ, Urbe S, Aniento F, Gruenberg J (1994) Vacuolar ATPase activity is required for endosomal carrier vesicle formation. J. Biol. Chem. 269:21-24

Diaz R, Wileman TE, Anderson SJ, Stahl P (1989) The use of permeabilized cells to study the ion requirements of receptor-ligand dissociation in endosomes. Biochem. J. 260:127-134

Freshney,R.I. (1994) In: Culture of Animal Cells: A Manual of Basic Technique. (3 eds) Wiley, New York, pp 317-318

Gerasimenko J, Tepikin A, Petersen O, Gerasimenko O (1998) Calcium uptake via endocytosis with rapid release from acidifying endosomes. Current Biology, 8:1335-1338

Mellman I (1996) Endocytosis and molecular sorting. Annu. Rev. Cell Dev. Biol.12:575-625

Mellman I, Fuchs R, Helenius A (1986) Acidification of the endocytic and exocytic pathways. Annu. Rev. Biochem. 55: 663-700

Mogami H, Tepikin AV, Petersen OH (1998) Termination of cytosolic Ca^{2+} signals: Ca^{2+} reuptake into intracellular stores is regulated by the free Ca^{2+} concentration in the store lumen. EMBO J. 17: 435-442

Parekh A, Penner R (1997) Store depletion and calcium influx. Physiol. Rev. 77:901-930

Petersen OH (1980) In: The electrophysiology of gland cells. Academic Press, London, New York, pp

Petersen OH (1992) Stimulus-secretion coupling: cytoplasmic and calcium signals and the control of ion channels in exocrine acinar cells. J. Physiol. 448: 1-51

Petersen OH, Petersen CCH, Kasai H (1994) Calcium and hormone action Annu.Rev. Physiol. 56:297-319

Pozzan T, Rizzuto R, Volpe P, Meldolesi J (1994) Molecular and cellular physiology of intracellular calcium stores. Physiol. Rev. 74:595-636

Saurin AJ, Hamlett J, Clague MJ, Pennington SR (1996) Inhibition of mitogen-induced DNA synthesis by bafilomycin A_1 in Swiss 3T3 fibroblasts. Biochem. J. 313:65-70

Smith C, Neher E (1997) Multiple forms of endocytosis in bovine adrenal chromaffin cells. J. Cell Biol. 139:885-894

Steinman RM, Mellman IS, Muller WA, Cohn ZA (1993) Endocytosis and the recycling of plasma membrane. J. Cell Biol. 96:1-27

Tepikin AV, Voronina SG, Gallacher DV, Petersen OH (1992 a) Acetylcholine-evoked increase in the cytoplasmic Ca^{2+} concentration and Ca^{2+} extrusion measured simultaneously in single mouse pancreatic acinar cells. J. Biol. Chem. 267:3569-3572

Tepikin AV, Voronina SG, Gallacher DV, Petersen OH (1992b) Pulsatile Ca^{2+} extrusion from single pancreatic acinar cells during receptor-activated cytosolic Ca^{2+} spiking. J. Biol. Chem. 267:14073-14076

Zen K, Biwersi J, Periasamy N, Verkman AS (1992) Second messengers regulate endosomal acidification in Swiss 3T3 fibroblasts. J. Cell Biol. 119:99-110

Part C

Misellaneous

Measuring Ca^{2+} Extrusion from Single Cells

PAVEL V. BELAN

Introduction

This chapter describes two techniques developed for measurements of calcium extrusion from single cells. The droplet technique was the first method that allowed direct visualization of calcium extrusion from single isolated mammalian cells (Tepikin et al. 1992). The droplet technique was initially used for measurements of calcium extrusion from gigantic neurons of snails (Tepikin et al. 1991). Later modifications of the technique made it possible to measure calcium extrusion that was mediated by calcium ATPases from pancreatic acinar cells (Tepikin et al. 1992a,b, Tepikin et al. 1994). It also allowed to monitor calcium efflux by exocytocis from submandibular acinar cells (Belan et al.1998). The main advantage of the droplet technique is that it allows to measure calcium extrusion which occurs during complex calcium reactions (e.g. calcium oscillations). The other technique described in this chapter employs heavy dextrans linked to calcium indicators in order to measure the distribution of calcium extrusion on the surface of singe isolated mammalian cells (Belan et al. 1996, Belan et al. 1998). This recently developed technique appears to be a useful tool for measurement of calcium extrusion from single cells. Later this technique was applied to measure calcium extrusion from isolated organelles: nuclear envelope (Gerasimenko et al. 1995) and isolated secretory granules (Gerasimenko et al. 1996, Nguyen et al. 1998).

Pavel V. Belan, The University of Liverpool, The Physiological Laboratory, P.O. Box 147, Liverpool, L69 3BX, UK, Bogomoletz Institute of Physiology, 4 Bogomoletz Street, Kiev-24, GSP 01024, Ukraine (*phone* +380-44-256-2428; *fax* +380-44-253-6458; *e-mail* pasha@serv.biph.kiev.ua)

Subprotocol 1
The Droplet Technique

Outline of the technique

The main idea of the droplet technique is to maintain a cell in a very small droplet; the volume of such droplet should be so small that calcium extrusion from even a single cell should be enough to produce measurable changes of extracellular calcium concentration. The schematic drawing of the droplet technique experiment is shown on Fig 1. A droplet is formed on the surface of a siliconised cover-glass and covered by mineral oil to prevent fast evaporation. The cell and extracellular solution are loaded with different types of calcium indicators. The optical properties of these indicators are selected to allow synchronous measurements of intracellular and exracellular calcium concentrations. The cell in the droplet can be stimulated by agonists applied via micropipettes that are inserted into the droplet. Estimation of the volume of the cell and the droplet allows an experimenter to calculate the changes of total calcium concentration in the cell and in the droplet.

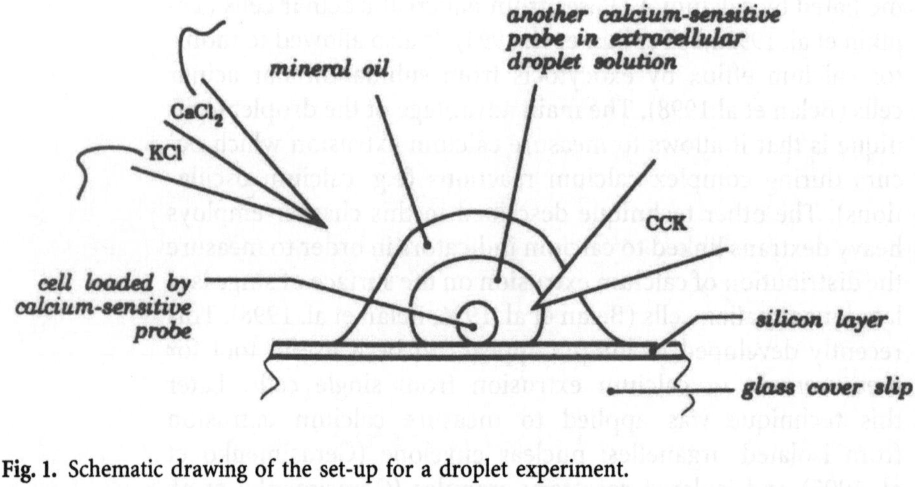

Fig. 1. Schematic drawing of the set-up for a droplet experiment.

Procedure

Preparation of cover-glasses

Cover-glasses mast be siliconised to prevent spreading of droplets.

1. Wash cover-glasses with detergent and rinse with distilled water. Wash these cover glasses with ethanol then rinse with distilled water and dry.
2. Immerse cleaned cover-glasses into Sigmacote solution (Sigma) for 5 minutes.
3. Remove cover-glasses from Sigmacote solution and allow the solution on the surface of cover-glasses to dry.
4. Rinse each cover-glass with distilled water (repeat rinsing with fresh distilled water 5 times).
5. Dry the cover-glasses and place them into a dust-free container for storage. Siliconised cover-glasses are reusable for a few weeks.

Making plastic pipettes

The cell containing droplet is formed on the surface of the cover-glass using two types of pipettes: coarse (with internal diameter of 200-400 μm) and fine (with internal diameter 50–100 μm). Pipettes were made from polyethylene tubing with internal diameter of 1-3 mm (Portex Limited, Kent, UK).

1. Cut polyethylene tube into 10 cm pieces.
2. Melt the central part of the tube over a heating element (e.g. platinum wire connected to current source and heated to red color). Slowly pull the ends of the tube apart to decrease the diameter of the tube in the central region.
3. Repeat the procedure 2-4 times until required diameter is obtained.

Droplet formation

The steps of the droplet formation are shown on Fig. 2.

1. Form a large drop (1-5 μl) of cell suspension in the central part of the siliconized cover-glass (Fig. 2a).
2. Immediately cover this drop with mineral oil to prevent evaporation.

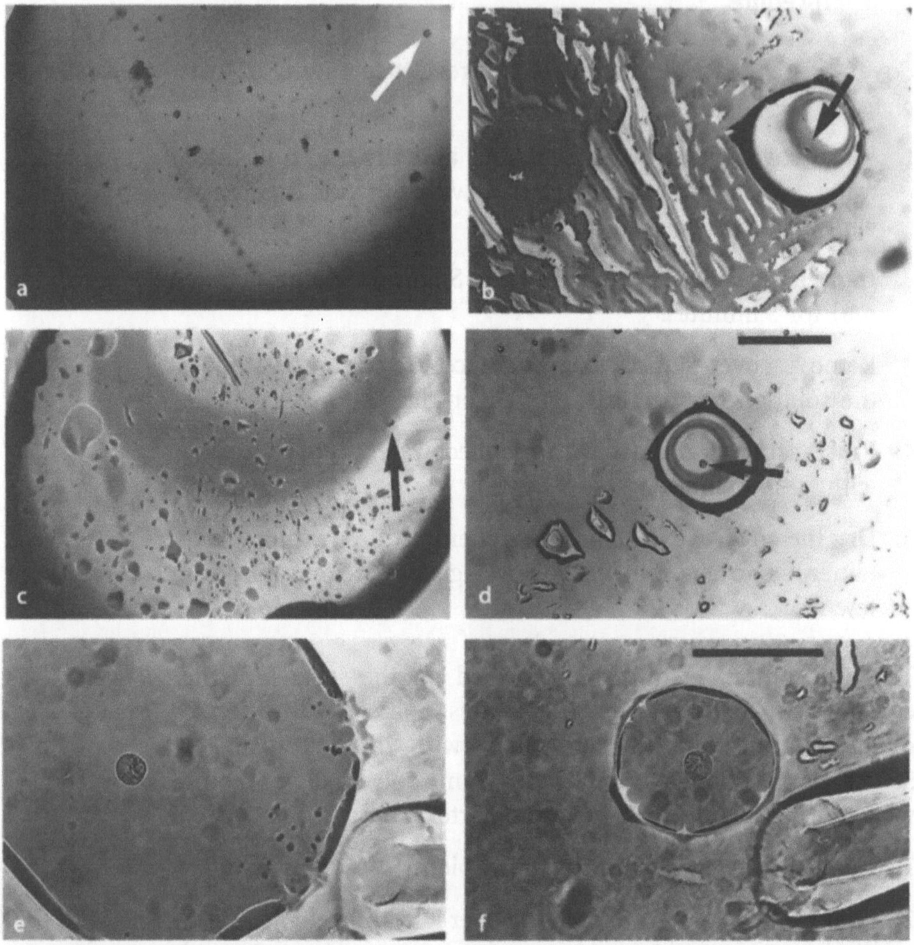

Fig. 2. Droplet formation: **a** - the initial drop contains many pancreatic acinar cells, the cell selected for the experiment is shown by a white arrow. **b** - other cells and excess of extracellular solution removed (cell selected for the experiment here and in C and D is shown by black arrow). **c** - injection of large amount of solution containing calcium indicator and known concentration of extracellular calcium. **d** - result of initial (without manipulator) removal of excess of extracellular indicator-containing solution. The length of the scale-bar for A,B,C,D equals 300 μm . **e**-droplet at the beginning of final adjustment, plastic pipette that was used for final steps of droplet formation is shown in the right part of this and next picture. **f**- completely formed cell-containing droplet. The length of the scale-bar for e and f represents 100 μm.

3. Select a cell suitable for the experiment.
4. Remove the other cells and most of the solution using a coarse pipette (Fig. 2b).
5. Infuse into the drop a large amount of solution with a known calcium concentration. This solution should also contain calcium indicator (Fig. 2c). Use a coarse pipette to infuse the solution. As a result of infusion the volume of the drop should increase by approximately 50-100 times (in comparison with the volume before infusion) (Fig. 2c).
6. Wait for 2-3 minutes then remove most of the solution using a coarse pipette (Fig. 2d).
7. When the size of the droplet is decreased to 300-400 μm change the pipette.
8. Use a fine pipette to further decrease the diameter of the droplet (Fig. 2 e,f). Use a manipulator to move the fine pipette (hydraulic manipulator MO 103, Narishigie was used in our experiments). The final diameter of the droplets in our experiments were usually in the range 80-140 μm.

Note: It is important to have the area of experimental setup dust free. Small particles of dust will distort the droplet, prevent reduction of the droplet's size, and will contaminate extracellular droplet solution with calcium.

Injection of substances into the droplet

To stimulate the cell, agonists will have to be delivered into the droplet solution. At the end of the experiment the extracellular calcium indicator should be saturated with calcium. Substances were injected into the droplet using micropipettes. Pressure injection, iontophoretic injection, and simple diffusion of substances were used in our experiments.

1. For pressure injection use pipettes of relatively large tip diameter (up to 2 μm in our experiments).
2. Siliconize external surface of pipettes by immersion into Sigmacote solution (Sigma) for 2-5 minutes. Dry and then rinse with distilled water. Siliconization prevents major distortion of the surface of the droplet by the pipette. In our experiments thapsigargin and cholecystokinin were frequently applied into the droplets by pressure injections.

3. Do not siliconize sharp (tip diameter less than 0.5 μm) microelectrodes for the iontophoretic application. Calcium and acetylcholine could easily be injected into the droplet by iontophoresis.

4. All substances could be delivered into the droplet by simple diffusion. Use this approach if you do not require to stimulate the cell at a precise moment in time.

Measurements of intracellular and extracellular calcium concentrations

The main combination of indicators used in our experiment was - fura-2 for intracellular calcium measurements and fluo-3 for monitoring extracellular calcium. An example of calcium concentration changes in extracellular droplet solution and inside the cell is shown in Fig. 3. Ideally a filter wheel with 3 excitation filters (e.g. 340nm, 380 nm and 490nm) should be used for this combination of indicators. Fura-2 does not fluoresce when excited at 490 nm (close to maximum of excitation of fluo-3). Fluorescence of fluo-3 at wavelengths 340 and 380 nm (wavelengths conventionally used to excite fura-2) is very small.

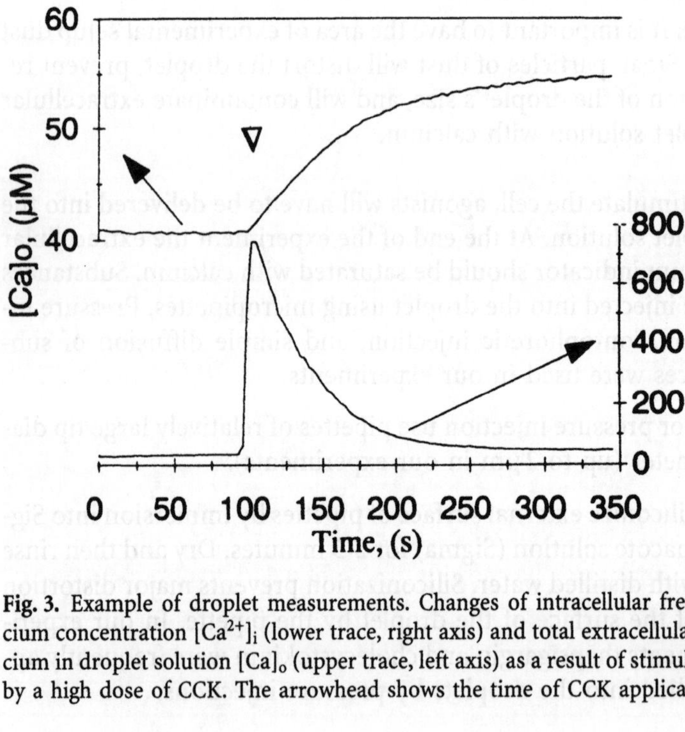

Fig. 3. Example of droplet measurements. Changes of intracellular free calcium concentration $[Ca^{2+}]_i$ (lower trace, right axis) and total extracellular calcium in droplet solution $[Ca]_o$ (upper trace, left axis) as a result of stimulation by a high dose of CCK. The arrowhead shows the time of CCK application.

This separation of excitation spectra allows synchronous measurements of intracellular and extracellular calcium. Fluo-3 behaves as true single wavelength indicator. The ratio of fluorescence of fluo-3 at 380 nm and 490 nm does not change with changes of calcium concentration. This allows to calculate and remove fluo-3 contribution from total fluorescence recorded at 380 and 340nm. The remaining fluorescence is exclusively due to fluorescence of fura-2.

Estimation of contribution of fluo-3 to fluorescence measured at 380 and 340 nm

1. Prepare cell-free droplets containing 50-100 μM of fluo-3 (concentration usually used in droplet experiments).
2. Measure intensity of fluorescence at 340,380 and 490 nm of excitation. The ratio of intensities of fluorescence is practically independent of calcium concentration in droplet but the fluorescence of fluo-3 is much brighter when indicator is saturated with calcium, therefore in these calibration experiments it is useful to add calcium to the droplet solution (e.g. to final concentration of 200 - 500 μm).
3. Calculate ratios: $\sigma_1 = F^{340}_{fluo}/F^{490}_{fluo}$ and $\sigma_2 = F^{380}_{fluo}/F^{490}_{fluo}$
4. Use these values to subtract fluo-3 contribution from total fluorescence excited at 340 and 380 nm in cell containing droplet experiments and in this way reveal fura-2 fluorescence in the cells:

   $F^{340}_{fura} = F^{340}_{total} - \sigma 1 * F^{490}_{fluo}$ similarly $F^{380}_{fura} = F^{380}_{total} - \sigma 2 * F^{490}_{fluo}$

Calibration and calculation of calcium concentration

1. At the end of the experiment inject calcium into the droplet solution to saturate extracellular fluo-3. In our experiments we used double-barrelled microelectrodes with one barrel filled with $CaCl_2$ (0.2 M) and the other with KCl (0.5 M) for calcium injection.
2. Calculate the total calcium concentration in extracellular droplet solution at each time-point of an experiment using the following formula:

$$[Ca]o = Kd(F - F_{min})/(F_{max} - F) + [fluo - 3](F - F_{min})/(F_{max} - F)$$

Where F_{min} is the fluorescence of the calcium free form of fluo-3 (fluorescence measured at 490 nm excitation), F_{max} is fluorescence of indicator saturated with calcium, F - fluorescence of the fluo-3 at a particular time-point measured in the droplet experiment.

3. Find fura-2 fluorescence at 380 nm (formula 2 above) by calculating and subtracting fluo-3 contribution from the total fluorescence at this particular wavelength.
4. Repeat the procedure for 340 nm excitation.
5. Find R values ($F^{340}{}_{fura}/F^{380}{}_{fura}$) for each time-point. Use conventional formula for ratiometric calcium indicators to calculate calcium concentrations (Grynkiewicz et al. 1985). Calibration of fura-2 (to determine R_{min}, R_{max} and β values) should be carried out independently on isolated cells placed in a perfused chamber.

Estimation of volumes of the cell and the droplet

To calculate the total calcium amount extruded by the cell it is necessary to calculate cell volume and the volume of the droplet. Modern confocal systems allow estimation of these volumes using standard 3D scanning software. In our experiments we measured cellular and droplet volumes using a calibrated ocular. The procedure is demonstrated in Fig. 4.

1. Measure the diameter of the cell using a calibrated ocular. Calculate the volume of the cell assuming that the cell is spherical. This approximation is of course suitable for only some cell types.
2. Take droplet into the fine plastic pipette (Fig. 4a,b).
3. Release the droplet over the surface of the cover-glass (Fig. 4c).
4. Wait until the droplet touches the surface of the cover-glass, at that moment measure the diameter of the droplet. Use this measurement to calculate the volume of the droplet.

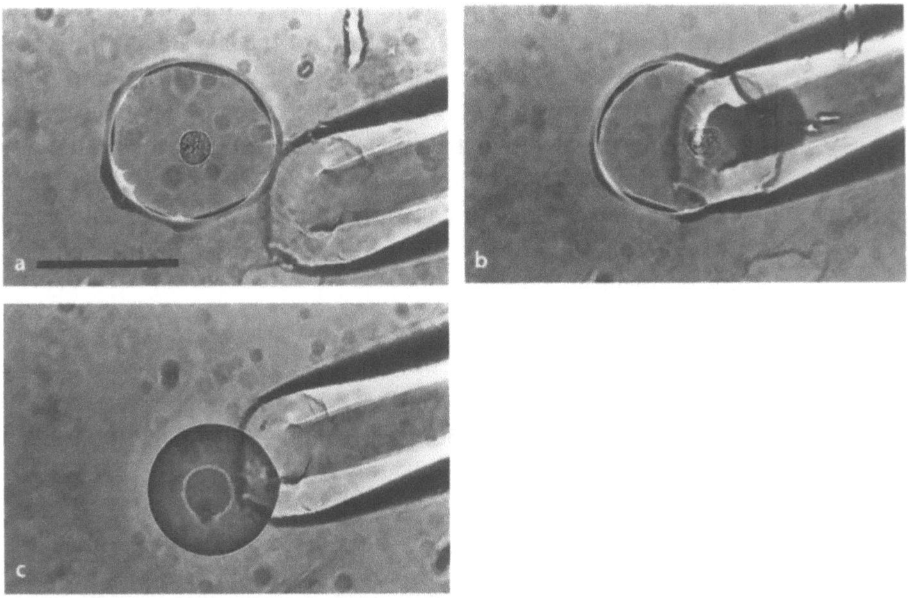

Fig. 4. Estimation of the droplet volume. Scale-bar corresponds to 100 µm. **a**-before the beginning of measurements, **b**- the droplet is taken into the pipette filled with oil, **c**- droplet released from the pipette on to the clean surface of the cover-glass. While it retains spherical shape (a few seconds), measurements of droplet diameter are performed.

Calculation of the total amount of calcium extruded from the cell and intracellular changes in the total calcium concentration

1. Calculate total amount of calcium extruded from the cell: $\Delta Ca_{cell} = \Delta[Ca]_{droplet} * V_{droplet}$ where ΔCa_{cell} is the change of the total amount of calcium in the cell, $\Delta[Ca]_{droplet}$ is the change of the total calcium concentration in the droplet, $V_{droplet}$ - the volume of the droplet. Changes of calcium concentration in the cell could be estimated by: $\Delta[Ca]_{cell} = \Delta Ca_{cell}/V_{cell}$.

Comments

Rhod-2, Calcium-green-5N, Mg-fura-2 (in these experiments fluo-3 was used as intracellular indicator) and Ca^{2+}- ion selective electrode were used to monitor the droplet's calcium concentration in our experiment. A modification of the droplet technique that allows simplification of the last stages of droplet formation was recently described by Schwiening (Schwiening 1996). The droplet technique was used to measure calcium dependency of calcium extrusion (Camello et al. 1996) and the rate of calcium secretion by secretory vesicles (Belan et al. 1998). The calcium dependency of calcium binding capacity of cytoplasm was measured recently using the droplet technique (unpublished results).

Subprotocol 2
Measurements of calcium extrusion using heavy dextrans

Outline

These are technically much simpler experiments than the measurements of extrusion using the droplet technique. The advantage of this type of experiment is that not just total intensity of extrusion but also the local intensities of calcium extrusion could be monitored at the single cell level. We found that the confocal microscope is an essential requirement for these types of measurements. Simple video imaging was not capable of resolving the details of calcium extrusion at cellular and subcellular levels. Calcium spreading from the place of extrusion was slowed down by means of a high-molecular weight Ca^{2+} buffer (calcium green-1 dextran, Mr 500,000) which, simultaneously was used as a calcium indicator. This allowed sufficient time to resolve the locations of calcium extrusion sites using confocal microscopy. Fig. 5 shows an example of such measurements. Distribution of intracellular calcium and calcium extrusion could be monitored simultaneously. The spectral properties of fura-red are considerably different from Calcium green-1-dextran, therefore, the intracellular and extracelluālr calcium concentrations could be monitored simultaneously (Fig. 6). Both fura-red and calcium green-1-dextran require visible wavelengths of light for excitation (e.g. 488 nm laser line) this means that UV range of wavelengths could be employed for photolyses of caged calcium. Combined experiments using cells loaded with and NP-EGTA (caged calcium) immersed in calcium green-1-dextran containing extracellular solution are possible (Fig. 7).

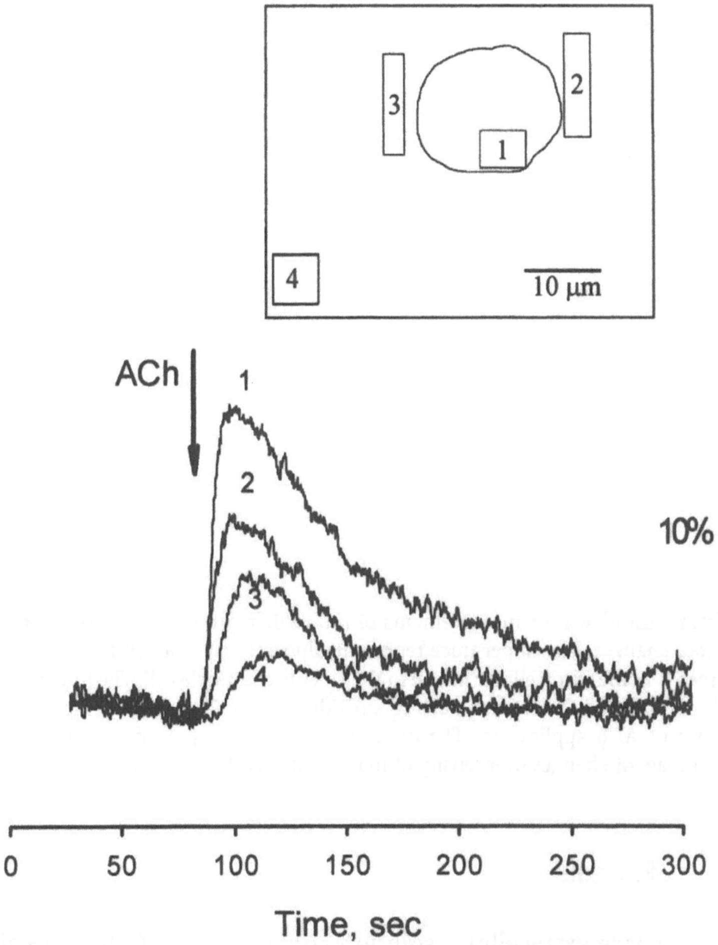

Fig. 5. Fluorescence changes measured in the vicinity of two connected pancreatic acinar cells stimulated by iontophoretic application of ACh. Extracellular solution contained 2.5 µM of calcium green-1 dextran (corresponds to 50 µM calcium green-1). The traces show calcium green-1 dextran fluorescence in different regions of the extracellular solution (fluorescence was recorded from a confocal section located slightly above the surface of the cells). The inset shows the boundary of the cells and the position of boxes. Numbers on the traces correspond to the number of the box from which fluorescence was recorded. The time of ACh application is indicated by the arrow. The application was 15 sec long. The vertical bar (here and in other figures) represents the ratio of the fluorescence changes divided by the initial level of the fluorescence.

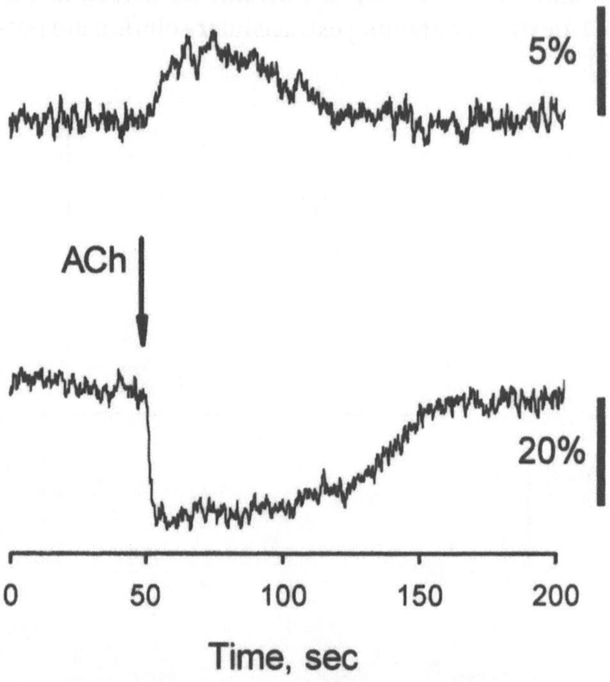

Fig. 6. Simultaneous measurements of intracellular and extracellular fluorescence changes. The upper trace represents fluorescence changes from calcium green-1 in the extracellular milieu in the vicinity of a single cell. The lower trace shows fura-red fluorescence changes inside the cell. The arrow indicates the onset of ACh application. The duration of the ACh application was 15 sec. The scale of changes of intensity of fluorescence is shown by the vertical bars.

Procedure

1. Prepare extracellular solution containing 30-100 μM of calcium sensitive dye linked to dextran. To increase sensitivity of the technique calcium green 1-dextran should be the only calcium buffer in the extracellular solution. In our experiments we used nominally calcium free solution.

Note: It is important to find out the number of molecules of indicator per molecule of dextran for the particular batch of the labeled dextran that you are planning to use in your experiments. Contact the supplier (The Molecular Probes, Eugene, USA) to obtain this information.

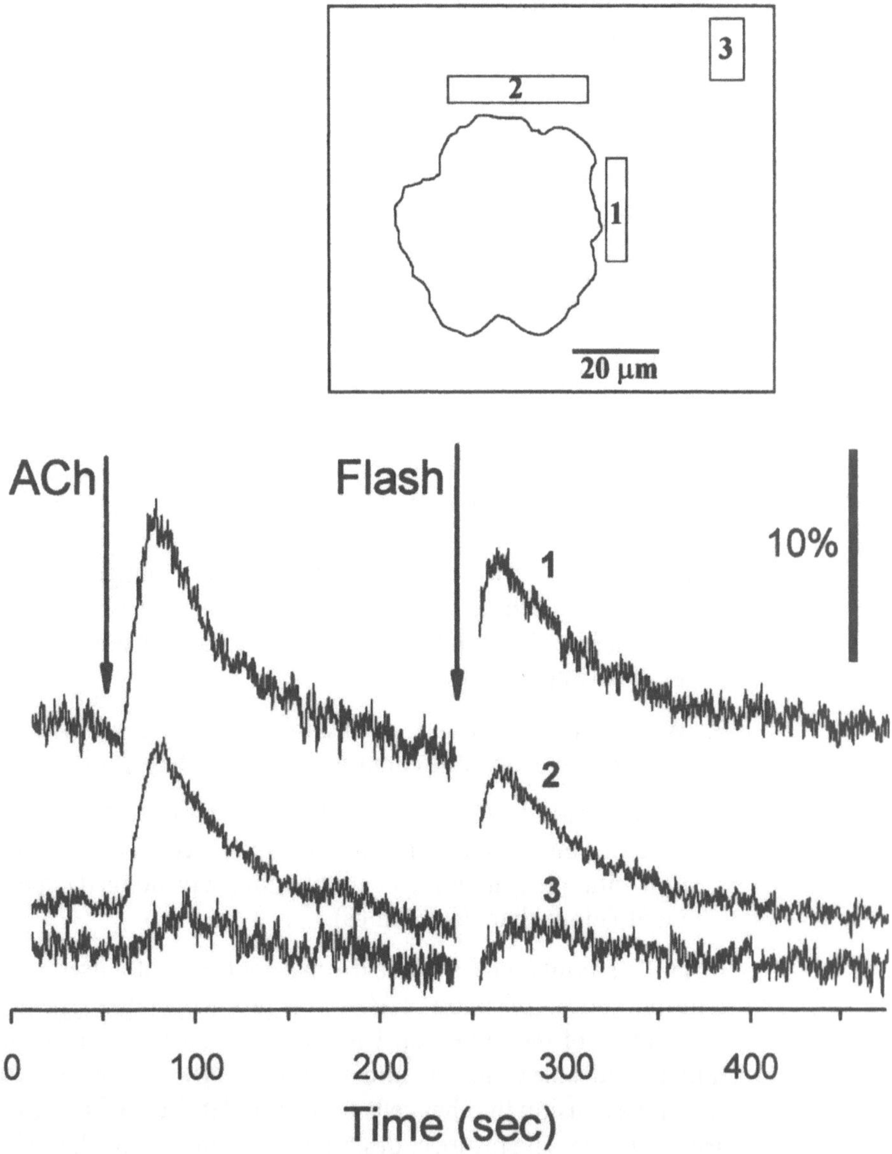

Fig. 7. Changes of calcium green-1 dextran fluorescence in the vicinity of a cell cluster (9 cells) induced by ACh application and by flash-photolysis of intracellular NP-EGTA. Arrows indicate time of ACh application and pulse of UV light. The traces show calcium green-1 dextran fluorescence in different regions of extracellular solution (fluorescence was recorded from a confocal section located slightly above the surface of the cells). The inset shows the boundary of the cells and the position of the boxes. Numbers on the traces correspond to the numbers on boxes from which fluorescence was recorded. The scale of changes of fluorescence intensity is shown by the vertical bar.

2. Test the solution for calcium contamination. To do this place a small drop (e.g.20 μl) of solution on cover-glass and measure the fluorescence. Saturate indicator with calcium by adding a small volume (e.g. 0.5 μl) of concentrated (e.g. 100mM) $CaCl_2$ containing solution and repeat the fluorescence measurements. Estimate the initial free calcium concentration in the solution using the conventional formula for single wavelength indicators. In our experiments free calcium concentration in calcium green-1 containing solution was usually in the range of 0.2 - 0.4 μM.

Note: In a few cases we found that calcium-green 1 dextran itself was considerably contaminated with calcium. If calcium contamination of extracellular solution is too high (e.g. $[Ca^{2+}]_o > 0.5$ μm), then the level of contamination should be reduced. This could be achieved using Calcium Sponge S. The concentration of calcium in calcium green-1 containing extracellular solution was reduced by incubation with Calcium Sponge S (40 mg per 0.5ml of solution). Calcium Sponge S was later removed by centrifugation. Calcium Sponge S was purchased from Molecular Probes (Eugen, USA).

3. Wash cells in calcium free solution and resuspend in a calcium-green 1 dextran containing calcium free solution.

4. Place 200 μl of cell suspension into the chamber. Mount the chamber on the stage of a confocal microscope. Cover solution with mineral oil if you are planning a prolonged experiment (more than 10 minutes).

Note: To stimulate cells in the droplet we frequently used local iontophoretic application of ACh. This protocol allows repeated stimulation of the same cell. The protocol also allows us to stimulate and study the calcium responses synchronously with calcium extrusion in other cells in the droplet. Labeled dextrans are expensive. The protocol described in section 4 together with local iontophoretic stimulation should allow the maximizing of cost efficiency of the experiments.

5. To stimulate cells with ACh using local application of the agonist. Load microelectrode (resistance 40-50 MΩ when filled with 1M KCl) with 10-20mM of ACh. Place another electrode (e.g. AgCl coated wire) into experimental chamber.

Insert microelectrode into the solution. Apply retaining current (5-10nA). Place the tip of the electrode at a distance of 50-100 μm from the group of cells that you are planning to stimulate. Apply injection current (10-100nA in our experiments) to stimulate the cell.

Note: Whilst calcium efflux from stimulated cells increases fluorescence in extracellular solution, large injection currents (more than 60nA) produce a small decrease of fluorescence in regions of the solution immediately adjacent (at a distance not more than 10 μm) to the tip of the microelectrode. This effect is present in cell free calcium green-1 dextran containing solution. The reason for this effect was not specifically investigated.

6. Once you have finished the experiments with a particular experimental chamber, you will need to saturate the extracellular indicator with calcium. Add a small volume (e.g. 1 μl) of a highly concentrated calcium solution (e.g. 1M) to your chamber to saturate extracellular calcium indicator. Measure the change in fluorescence. Use this value to calculate changes of free calcium concentrations in regions that surround stimulated cells during experiments.

Comments

The technique employing heavy dextrans could potentially be used to measure extrusion of other types of ions (e.g. H^+), for which heavy dextrans bound to appropriate indicators are available.

The process of calcium diffusion in a dextran bound to calcium indicator can be calculated (e.g. equation 11 in Belan et al. 1996). We used these calculations only at the time of the initial development of the technique when it was necessary to test the feasibility of this method of measurements. In the majority of the types of experiments the concentration of calcium green 1 dextran was adjusted empirically. Pancreatic acinar cells could generate a calcium efflux of approximately 2×10^{-17}mol s^{-1}. If the cells that you are planning to study could produce equal or more powerful calcium extrusions, then the technique employing heavy dextrans could be used.

References

Belan PV, Gerasimanko OV, Berry D, Saftenku E, Petersen OH, Tepikin AV (1996) A new technique for assessing the microscopic distribution of cellular calcium exit sites. Pflugers Arch 433: 200 - 208.

Belan PV, Gerasimenko OV, Tepikin AV, Petersen OH (1996) Localization of Ca^{2+} extrusion sites in pancreatic acinar cells. J Biol Chem 271: 7615 - 7619.

Belan P, Gerasimenko O, Petersen OH, Tepikin AV (1997) Distrebution of Ca^{2+} extrusion sites on the mouse pancreatic acinar cell surface. Cell Calcium 22: 5-10.

Belan P, Gerasimenko O, Petersen OH, Tepikin AV (1998) Isoproterenol evokes extracellular Ca^{2+} spikes due to secretory events in salivary gland cells. J Biol Chem 273: 4106-4111.

Camello P, Gardner J, Petersen OH, Tepikin AV (1996) Calcium dependence of calcium extrusion and calcium uptake in pancreatic acinar cells. J Physiol 490: 585 - 593.

Gerasimenko OV, Gerasimenko JV, Tepikin AV, Petersen OH (1995) ATP-Dependent Accumulation and Inositol Trisphosphate or Cyclic ADP-ribose Mediated Release of Ca^{2+} from the Nuclear Envelope. Cell 80: 439 - 444.

Gerasimenko OV, Gerasimenko JV, Belan PV, Petersen OH (1996) Inositol trisphosphate and cyclic ADP-ribose - mediated release of Ca^{2+} from single isolated pancreatic zymogen granules. Cell 84: 473-480.

Grynkiewicz G, Poenie M, Tsien RY (1985) A new generation of Ca^{2+} indicators with greatly improved fluorescence properties. J Biol Chem 260: 3440-3450.

Nguyen T, Wei-Chun Chin, Verdugo P (1998) Role of Ca^{2+}/K^{+} ion exchange in intracellular storage and release of Ca^{2+}. Nature 395: 908-912.

Schwiening CJ (1996) Drawing on glass under oil: An easy way of forming small droplets around single cells. J Physiol 493P: P4-P5.

Tepikin AV, Kostyuk PG, Snitsarev VA, Belan PV (1991) Extrusion of calcium from a single isolated neuron of the snail Helix Pomatia.. J Membrane Biol 123: 43-47.

Tepikin AV, Voronina SG, Gallacher DV, Petersen OH (1992a) Acetylcholine-evoked increase in cytoplasmic Ca^{2+} concentration and Ca^{2+} extrusion measured simultaneously in single mouse pancreatic acinar cells. J Biol Chem 267: 3569-3572.

Tepikin AV, Voronina SG, Gallacher DV, Petersen OH (1992b) Pulsatile Ca^{2+} extrusion from single cells during receptor -activated cytosolic Ca^{2+} spiking. J Biol Chem 267: 14073-14076.

Tepikin AV, Llopis J, Snitsarev VA, Gallacher DV, Petersen OH (1994) The droplet technique: measurement of calcium extrusion from single isolated mammalian cells. Pflugers Arch 428: 664-670.

Tracking Calmodulin Inside Cells

MADELEINE CRASKE

Introduction

Calmodulin, originally identified by two independent laboratories as an activator of brain cyclic nucleotide phosphodiesterase (Cheung, 1970; Kakiuchi et al., 1970), has since become widely recognized as one of the most important and abundant proteins in eukaryotic cells. Calmodulin is now understood to be one of the major intracellular target proteins for calcium and many important enzymes and membrane transport systems are directly regulated by calmodulin, including calcium calmodulin dependent kinases and phosphatases, adenylate cyclase, cyclic nucleotide phosphodiesterase and calcium ATPases.

As some of the most important targets for calmodulin are located in the nucleus, researchers in this field have recently been focusing on calcium dependent translocation of calmodulin into this organelle. Movement of calmodulin may enable it to act as a mobile calcium buffer and sensor throughout the cell, and in addition, the translocation of calmodulin will determine its availability for binding targets in specific areas of the cell. The translocation of calmodulin into the nucleus has been demonstrated in several systems including adrenal cells (Harper et al., 1980), liver cells (Pujol et al., 1989), and smooth muscle cells (Luby-Phelps et al., 1995), and the mechanism of its import has been investigated (Pruchy et al., 1994; Liao et al., 1999). One of the main functions of calmodulin in the nucleus is the regulation of gene expression (Bachs et al., 1994), which

Madeleine Craske, Stanford University, Department of Molecular Pharmacology, CCSR Building, Room 3230, 269 Campus Drive, Stanford, CA 94305, USA (*phone* 650-725-7445; *fax* 650-725-2952; *e-mail* craskem@stanford.edu)

has been shown by various groups to be accomplished via CaM kinase triggered phosphorylation of a transcription factor, called cyclic AMP response element binding protein (CREB) (Dash et al., 1991; Matthews et al., 1994; Mayford et al., 1995; Bito et al., 1997; Finkbeiner & Greenberg, 1997). Recently Deisseroth and co-workers have succeeded in correlating the translocation of calmodulin into the nucleus with the phosphorylation of CREB in hippocampal neurons, thus helping to increase understanding of the way in which calcium signals in the cytosol may influence events occurring in the nucleus (Deisseroth et al., 1998).

The objective of the procedure described in the first section of this chapter is to introduce fluorescently labelled calmodulin into the cell, and then observe the intracellular distribution of calmodulin following various cell stimulation protocols. For these experiments we use calmodulin conjugated to a derivative of fluorescein (5-(4,6-dichlorotriazinyl)-amino-fluorescein-calmodulin, abbreviated to DTAF-CaM) which does not change its fluorescence when calcium binds. The main finding of this work is that a submaximal dose of agonist, which produces oscillating calcium signals, also results in oscillations of calmodulin movement with different patterns in different areas of the cell.

The final section of the chapter will deal with experiments that can be performed using TA-CaM (2-chloro-(ε-amino-Lys_{75})-(6-(4-N,N-diethylamino-phenyl)-1,3,5-triazin-4-yl)-calmodulin. The fluorescence of this form of calmodulin increases 2-4 fold upon calcium binding (Craske et al., 1999), therefore this compound can be employed to monitor the calcium calmodulin binding reaction in different regions of the cell.

Materials

Equipment

- laser scanning confocal microscope and ODYSSEY software (Noran Instruments, Middleton, WI) consisting of an inverted microscope with x40 NA 1.3 oil objective (Nikon)
- patch clamp amplifier (LM EPC7, List Electronics, Darmstadt, Germany)
- VBF/4 dual channel variable low pass filter (Kemo Ltd. England)

- Cambridge Electronics design 1401 interface (Cambridge, UK)
- Narishige micromanipulator (Japan)
- Leitz manipulator (Leitz Instruments, Germany)
- Compressed air vibration isolated table
- DMZ Universal pipette puller (Zeitz-Instrumente, Germany)
- Assistent microhaematocrit tubing (Assistant, Denmark)
- Perfusion system for solution changes

The cell perfusion system needs to be capable of allowing rapid changes of the cell bath fluid. Ideally it will have an inflow tube that can be positioned reasonably close to the calmodulin-loaded cell (within 100-300 μm) so that application of an agonist is fast and efficient. However, suction of waste fluid should be gentle to avoid disturbing or washing away the cells. The perfusion system may consist of a series of 20-50 ml syringe tubes for the control (Na^+ HEPES alone) and test solutions (Na^+ HEPES with agonist) with thin (1.5 mm O.D., 0.5 mm I.D.) polythene tubing taking solution to the cell bath via hydrostatic pressure. Alternatively, a peristaltic pump may be used to deliver solution to the bath. The advantage of the pump is that once the required speed has been determined, flow of solution to the cell bath should occur at a continuous, fixed rate.

Standard Extracellular Solution - Na^+HEPES

140 mM	NaCl
4.7 mM	KCl
1.13 mM	$MgCl_2$
1 mM	$CaCl_2$
10 mM	Glucose
10 mM	HEPES

The final osmolarity of the solution should be checked with an osmometer to be 290 ± 5m osM and pH should be adjusted to 7.2 with sodium hydroxide solution (NaOH). For experiments requiring $0Ca^{2+}$ in the external medium, $CaCl_2$ should be omitted from the solution and any remaining calcium chelated by adding 50 μM EGTA.

Standard Internal Medium - K^+ HEPES

140 mM	KCl
1.13 mM	$MgCl_2$
10 mM	HEPES
10 mM	Glucose
1 mM	ATP
0.1 mM	EGTA

The desired osmolarity of this solution is 290 ± 5m osM and pH should be adjusted to 7.2 using potassium hydroxide (KOH). This solution is used to dilute the TA-CaM and DTAF-CaM to the final working concentration of 100μM, and 20, 50 or 100μM, respectively. Therefore, it is good practice to filter K^+ HEPES before the dilution step with Acrodisc 0.2 μM filters so that any particles are removed, avoiding blockage of the pipette tip.

Subprotocol 1
Injection of Labelled Calmodulin into Cells

We have found that it is possible to utilize the whole cell patch clamp configuration to introduce fluorescently labelled molecules into the cytosol of pancreatic acinar cells. In order to test whether the protocol would be suitable for a 17 kD protein like calmodulin, we began with a much larger molecule, 70 kD fluorescently linked dextran. Within 3 minutes fluorescent dextran diffused throughout the entire cell, although the fluorescent signal from the nucleus was low, due to the fact that passive diffusion through nuclear pore complexes is limited for such a large molecule. It is possible that this technique could lend itself to the introduction of a variety of labelled probes such as small proteins, but its adaptability is likely to become limited by the size of molecule, i.e. very large molecules may be unable to diffuse through the pipette tip, diffusion will become very slow or the molecule may be unable to penetrate all regions of the cell.

Procedure

Preparation of pancreatic acinar cells and injection of DTAF-CaM

1. Isolate single cells and doublets from a fresh pancreas by injecting the tissue with 100-200 units collagenase (in calcium containing Na^+ HEPES) and incubate at 37 °C in an agitating water bath. Shake tissue to liberate cells.

2. Remove the collagenase from the resulting cell suspension by centrifugation and then re-suspend cells in collagenase-free solution.

3. A small quantity of the cell suspension (~200-400 µl) should be placed in a Perspex chamber with a disposable glass coverslip forming the base. The coverslip should be sealed to the bath using grease. The chamber should have dimensions suitable for inclusion of the patch clamp pipette, the reference electrode and the cell perfusion system.

4. Wait a few minutes for cells to adhere to the coverslip and then begin perfusing the cells with standard Na^+ HEPES containing solution.

5. A DMZ universal puller may be used to manufacture fire polished patch clamp pipettes with a tip diameter of 1 µM and a resistance of between 1.5 and 3 MΩ. A fine micropipette (Eppendorf) can be used to load 3 µl of TA- or DTAF-CaM as close as possible to the tip of the glass pipette.

6. Gently flick or tap the pipette to dispel any air bubbles and to ensure that the calmodulin solution has reached the very end of the pipette. Pipettes should then be back-filled with filtered K^+ HEPES solution.

7. Position the pipette so that it touches the basal membrane of the cell. Apply gentle negative pressure to obtain a giga seal between membrane and glass pipette.

8. Begin confocal imaging. For DTAF-CAM, the excitation and emission wavelengths are 488 nm and above 515 nm respectively, whereas for TA-CaM they are 364 nm and above 400 nm respectively. We use a linear colour scale to colour code the intensity of the calmodulin fluorescence in different cellular compartments.

9. Destroy the region of membrane under the pipette by brief application of suction to achieve the whole cell patch clamp technique and observe the diffusion of calmodulin into the cell.

10. At the end of the experiment take a bright field/transmitted light picture of the cell(s) so that the position of the secretory granules and the borders between cells in a cluster can be checked during data analysis.

Results

The result of a typical experiment where DTAF-CaM was injected into the cell by the whole cell patch clamp configuration is shown in Figure 1.

A linear colour scale was used to produce this figure, where warm colours such as red correspond to a higher concentration of DTAF-CaM. Figure 1A shows that the accumulation of 17 kD DTAF-CaM in the cell is fast and linear for the first 2 minutes, before gradually reaching a plateau after 6-7 minutes. The level of autofluorescence measured from a neighbouring cell remains

→

Fig. 1. The injection of DTAF-CaM into a mouse pancreatic acinar cell using the whole cell patch clamp configuration

A. The kinetics of DTAF-CaM diffusion from a patch pipette into the cytosol of a cell. The blue trace (labelled as cell) represents the intensity of DTAF-CaM fluorescence measured from the whole cell against time, while the black trace (labelled *af*) shows the level of autofluorescence from a neighbouring cell in the cluster. A giga seal between a DTAF-CaM loaded pipette and the cell membrane was achieved at time zero, and confocal imaging of the cell began. After approximately 17 seconds the whole cell configuration was obtained and DTAF-CaM accumulated rapidly in the cell.

B. Series of confocal images showing the diffusion of DTAF-CaM from the patch pipette into the cytosol of a cell.

a). The patch pipette containing 50 μM DTAF-CaM can be seen as a yellow-green area in the left half of the image. The cell to be loaded with DTAF-CaM can be seen on the right. At this stage, the cell-attached configuration had been achieved, but the cell membrane remained intact.

b). Between image a) and b), a period of approximately 17 seconds, the area of membrane under the patch pipette was destroyed.

c) to h). DTAF-CaM diffuses into the cell, as indicated by the progressive increase in the intensity of the yellow, and then red, colour of the cell.

i). Nonconfocal transmitted light picture of the cluster of cells selected.

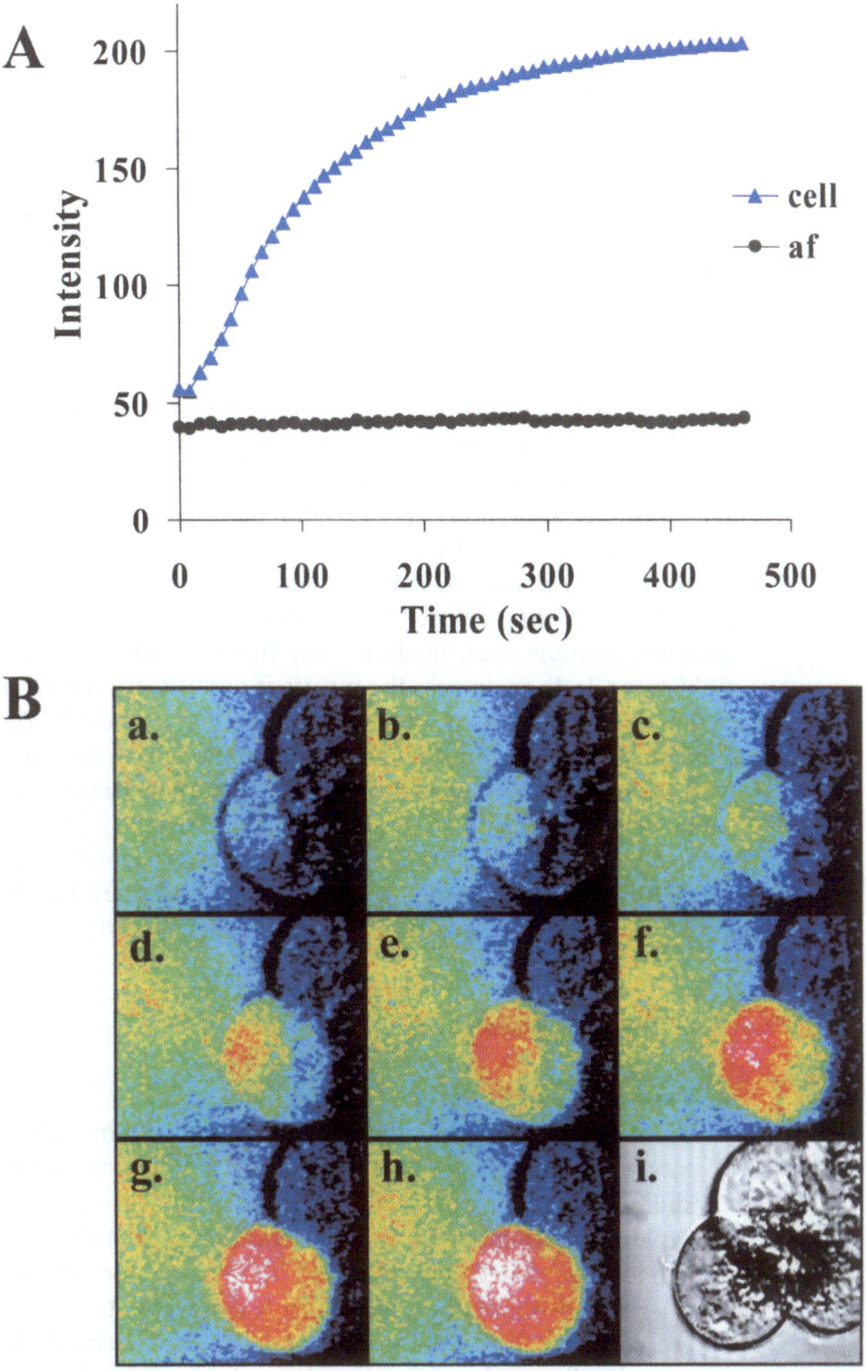
A
200
150
100
50
0
Intensity
cell
af
0
100
200
300
400
500
Time (sec)
B
a.
b.
c.
d.
e.
f.
g.
h.
i.

constant throughout the experiment showing that DTAF-CaM is unable to cross gap junctions. This also shows that the compound is stable - there is no dissociation of the fluorescein molecule, which may be small enough on its own to diffuse into neighbouring cells in the cluster. The fluorescence measured from the target cell may be higher than that recorded from the neighbour cell at the start of the experiment due to fluorescence glare from the cell-attached DTAF-CaM containing pipette. Figure 1B shows a selection of eight confocal images of the cell taken at different times during the course of the experiment. In Fig. 1B a) the pipette is attached to the cell and fluorescence from the DTAF-CaM it contains is seen as a diffuse yellow-green region to the left of the cell. At this point, the membrane of the target cell is intact and so the cell appears dark blue. During the 17 second time interval between image a) and image b) the cell membrane is broken and DTAF-CaM rapidly accumulates in the cytosol under the region of the membrane where the pipette is attached. By image h), 2 minutes and 15 seconds after image b), DTAF-CaM has diffused throughout the cell including the secretory granule area, furthest away from the site of DTAF-CaM entry. In these images the pipette is out of focus therefore its fluorescence intensity appears less than that of the cell. The final image is a nonconfocal light picture of the cell cluster taken after the removal of the pipette at the end of the experiment. The target cell, loaded with DTAF-CaM for approximately 7.5 minutes, appeared to suffer no detrimental effects as a result of this procedure. The membrane quickly resealed, and there was no swelling, or loss of the typical polarized cell structure.

Troubleshooting

- The patch pipette vibrates constantly:
  To minimise vibrations the microscope and the manipulators for the head stage of the patch clamp amplifier should be situated on an air isolation table. It is advisable before each experiment to check that the isolation table is sufficiently and evenly inflated at each corner. Ideally a remote confocal laser source should be used which is connected to the microscope via fibre optic cables and ventilators should be placed well away from the air isolation table. This is because

in our experience, electric cooling fans in the visible laser and the water circulation in the cooling system of the UV laser are considerable sources of vibration that prevent long term stability of the patch clamp recordings.

- The electrical recordings are very noisy:Enclose the table and microscope in a Faraday cage.
- The patch pipette resistance is smaller, or larger, than anticipated before making contact with the cell membrane:
  A low resistance indicates that the pipette tip may have been broken during the process of lowering it into the bath, while a high resistance may mean that cell debris in the bath solution has become attached to the tip. One way to avoid colliding with debris when moving through the bath is to apply constant positive pressure until the pipette is within a few micrometres of the cell surface.
- The cell swells rapidly after formation of whole cell configuration:
  Check osmolarity of the calmodulin solution - ensure that it is in the region of 290±5m osM. Cell swelling can sometimes be combated by the addition of 20 mM sucrose to the Na^+ HEPES perfusate. Increasing the osmolarity of the external medium reduces the tendency of cells to take up extra fluid.

Subprotocol 2
Monitoring Ca^{2+}-Induced Intracellular Calmodulin Movements

Procedure

The intracellular distribution of DTAF-CaM at rest is uniform throughout the basal non-nuclear region of the cell, but tends to be more concentrated in the nuclear compartment. This has been observed previously for the distribution of other fluorescent probes and probably results from the absence of membrane-bound organelles in the nucleoplasm. The basal non-nuclear region is densely packed with endoplasmic reticulum which effectively decreases the 'free cytosol' in this region and so represents a smaller volume for DTAF-CaM to occupy. In order to study the effect of a rise in intracellular calcium concentration on the distribution of fluorescent calmodulin, follow steps 1-7 from Subprotocol 1 of this chapter and then proceed as follows:

1. After formation of the giga seal, rupture the membrane under the pipette tip to obtain the whole cell configuration.

2. Wait 3-5 minutes to allow diffusion of DTAF-CaM into the cytosol.

3. To avoid the continuous flux of labelled calmodulin into the cell, which could mask the redistribution process, the pipette should be gently withdrawn before the start of confocal measurements. To achieve this aim without dislodging the cell from the coverslip, it is necessary to make only very small adjustments of the micromanipulator controlling the amplifier head stage/pipette position. The easiest method for removal involves alternating between a) drawing the pipette slowly away from the cell (for only a couple of micrometers) and b) moving the pipette in an upward direction for a similar distance, until the pipette is free.

4. Begin confocal imaging in time series mode with a suitable number of images and time resolution (for example, 100 images with an interval of 15 seconds between images resulting in a total recording time of approximately 25 minutes).

5. Record the resting distribution of DTAF-CaM for 1-2 minutes while perfusing the cell bath with normal Na^+ HEPES.

6. After acquiring the desired number of images of the cell under resting conditions, switch off the standard Na^+ HEPES perfusion and begin perfusing with Na^+ HEPES containing a supramaximal dose of agonist (1 μM ACh or 1nM CCK).

7. The location of the nucleus can be verified at the end of the experiment by short incubation of the cell with 50 μM of the DNA stain Hoechst 33342. After waiting 2-3 minutes for the Hoechst 33342 to enter the cell, it can be excited using the 364 nm laser line and emission should be recorded above 400 nm.

8. Obtain a transmitted light picture of the cells so that their orientation and granular positions can be checked during data analysis.

9. As a control experiment to test the specificity of CaM translocation, cells can be loaded with fluorescein-labelled dextran (M_r 10,000) via a patch pipette and the whole cell configuration. The cells should then be stimulated by 1nM CCK, or 1 μM ACh, to demonstrate that the distribution of this compound, whose size is similar to that of calmodulin, but which has no biological activity, does not change following an elevation of $[Ca^{2+}]_i$.

10. The data can be analyzed by measuring the fluorescence intensity changes in different regions of interest against time, and traces may be corrected for photobleaching using parts of the traces obtained before agonist application.

Results

The effect of applying a supramaximal dose of the agonist CCK on the movement of DTAF-CaM introduced into one of the cells of a doublet is shown in Figure 2.

DTAF-CaM fluorescence was monitored in the basal nonnuclear region (green trace), in the nucleus (blue trace) and in the apical secretory granule (SG) region (red trace). CCK evoked a steep rise in the fluorescence from the SG region followed by a more substantial but slower increase in the nucleus. These changes were accompanied by a corresponding decrease in DTAF-CaM concentration in the basal nonnuclear region. In all experiments the total averaged cellular fluorescence (when corrected for photobleaching) remained unchanged, indicating that the local changes are due to redistribution of calmodulin. Similar results are obtained when DTAF-CaM loaded cells are stimulated by 1 μM ACh, or when a calcium ionophore such as ionomycin is used. The fact that ionomycin produces the same pattern of redistribution as ACh or CCK confirms that these changes are due to a rise in intracellular calcium and not to some other biochemical pathway activated by these agonists. As the agonist-evoked rise in CaM concentration is initially much steeper in the SG region than in the nucleus, one potentially interesting protocol would be to apply shorter pulses of stimulation to determine whether CaM translocation into the SG region is favoured. The result of applying a relatively short (15-25 second) pulse of ACh (1μM) is shown in Fig. 3.

A brief but clearly resolved transient rise in CaM concentration occurs in the SG region and in 5 out of 11 such experiments a small nuclear rise was seen, but these nuclear changes were always much smaller than the SG response. Another variation on the protocol described above is to change the concentration of agonist used to stimulate the cells. In pancreatic acinar cells, low doses of the hormone CCK are known to evoke an oscillating Ca^{2+} signal. At 50 pM, CCK produces a repetitive pattern of global Ca^{2+} spikes rising from the basal level, with each transient lasting about 20-50 seconds. Typically, the first spike is two to four times longer than each of the following spikes (Petersen et al., 1991; Thorn et al. 1993). A typical example of the effect of 50 pM CCK on calcium levels in a Fura-2AM loaded cell is shown in Fig. 4A.

Fig. 2. DTAF-CaM undergoes redistribution following the application of CCK. **(A)** Relative changes in fluorescence is recorded in three regions of interest. The presence of CCK in the extracellular medium is indicated by the bar at the top of the graph. The CCK evoked a rapid transient increase in the fluorescence signal from the SG region (red trace) which is followed by a delayed, but stronger, signal from the nucleus. These effects are accompanied by a decrease in the fluorescence intensity from the basal nonnuclear area (green trace). The traces were corrected for photobleaching. Arrowheads at the top of the graph show the time points at which the five confocal images *(b-f)* in B were obtained. **(B,** ***a*****)** Nonconfocal transmitted light picture of the pancreatic acinar cell doublet selected. DTAF-CaM was loaded into the cell on the left via a patch pipette. The boxes over the nucleus (blue), SG region (red) and basal nonnuclear area (green) show where the fluorescence intensity changes were measured. Scale bar = 10 μM. **(B,** ***b-f*****)** Colour coded images of fluorescence intensity (linear scale). *(b)* Image taken prior to the application of CCK. The injected cell is brightly stained with DTAF-CaM, which is present throughout the cytosol, but is more concentrated in the area corresponding to the nucleus. *(c-f)* Series of images revealing the intracellular redistribution of DTAF-CaM evoked by CCK. *(c)* Before the peak response to CCK, DTAF-CaM appears to be accumulating initially in the SG region (due to the increased intensity of yellow in this area when compared to b). *(d)* At the peak of the response the nuclear area is bright red, indicating a substantial accumulation of DTAF-CaM in this region. Note that the nonnuclear region is now blue rather than green, suggesting that the concentration of DTAF-CaM in this area has decreased. *(e)* The fluorescence signal from the SG region is now reduced significantly (compared with d). *(f)* The distribution of fluorescence is similar to that observed before stimulation *(b)*. The overall intensity is somewhat lower than in b as a result of photobleaching. This figure has been reproduced with kind permission of PNAS (Craske et al., 1999).

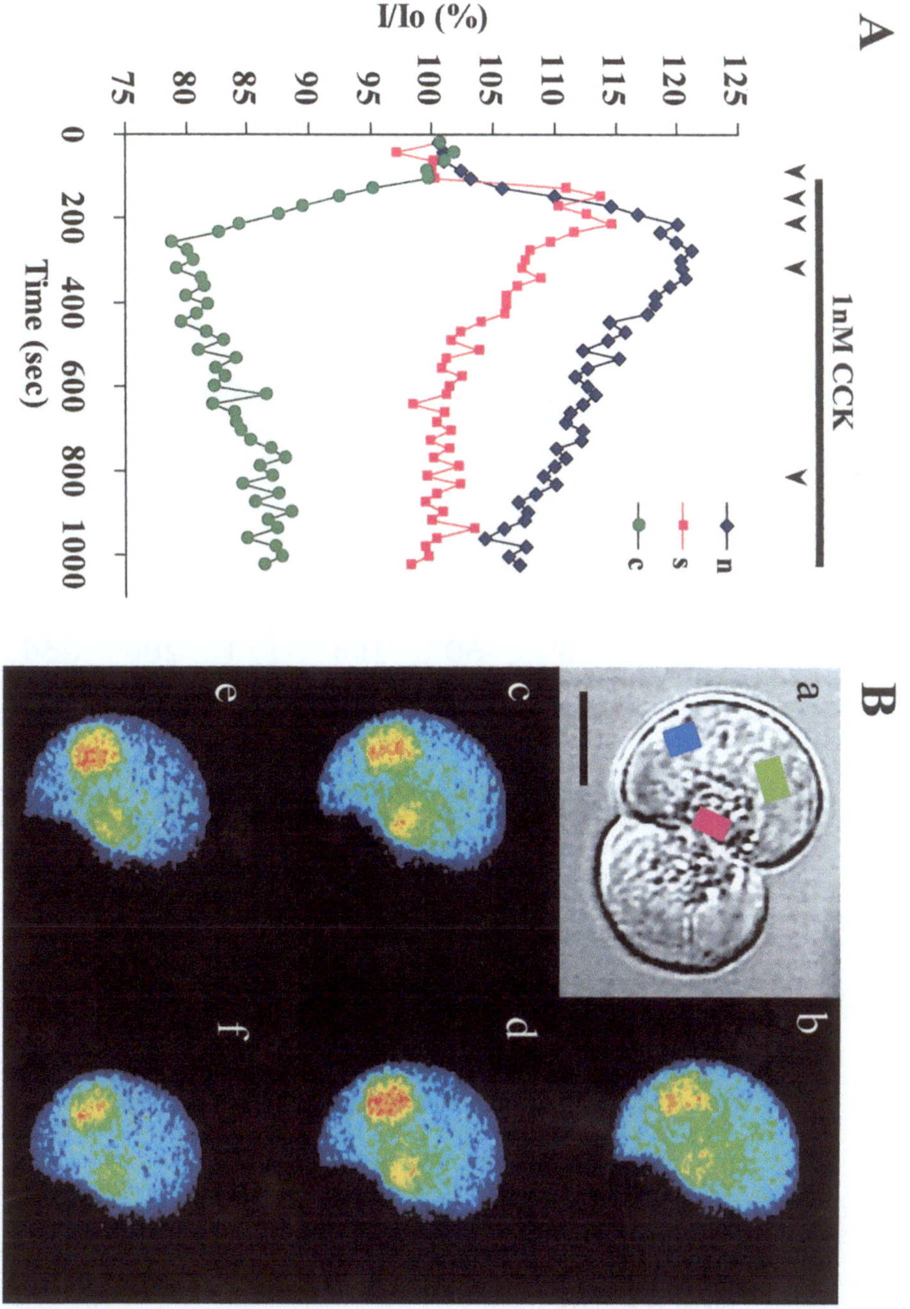
A
1nM CCK
I/Io (%)
125
120
115
110
105
100
95
90
85
80
75
0
200
400
600
800
1000
Time (sec)
n
s
c
B
a
b
c
d
e
f

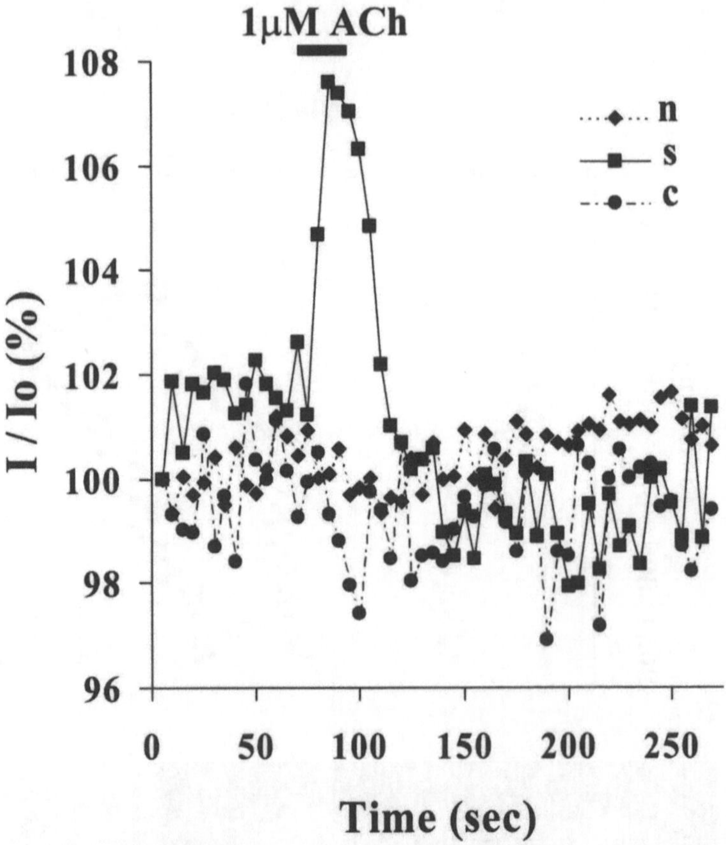

Fig. 3. Short application of ACh (indicated by the bar) triggers a transient rise in CaM concentration in the SG region, but has little or no effct on nuclear CaM concentration. Relative changes of DTAF-CaM fluorescence in the nucleus (diamond symbol, dotted line), SG region (square symbol, solid trace) and basal nonnuclear region (circular symbol, dashed line) are shown. This figure has been adapted from Craske at al., 1999.

→

Fig. 4. CCK (50 pM) induces uniform $[Ca^{2+}]_i$ oscillations, but different types of oscillations of CaM concentration in different regions of pancreatic acinar cells. The point of agonist application is shown by the short vertical bars on the *x*-axes and continues for the period indicated by the horizontal bars at the top of the graph. The $[Ca^{2+}]_i$ (measured with Fura-2) and the CaM oscillations were monitored in separate cells. **(A)** Calcium oscillations in three regions of interest: nucleus (diamond symbol, dotted line), SG region (square symbol, solid trace) and basal nonnuclear region (circular symbol, dashed line). **(B)** Oscillations of calmodulin concentration (oscillations of the intensity of DTAF-CaM fluorescence) in the three regions of interest. This figure has been adapted from Craske et al., 1999.

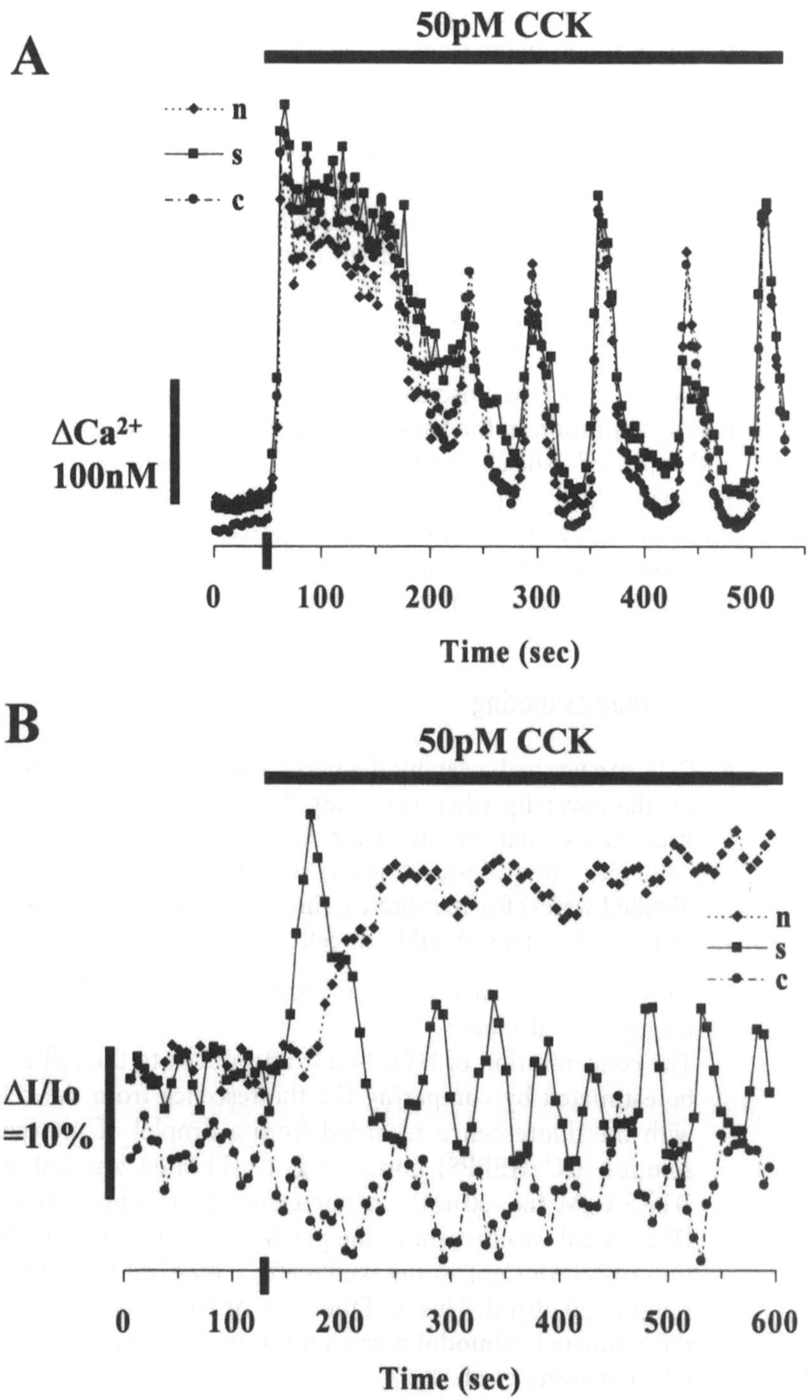
A
50pM CCK
n
s
c
ΔCa2+
100nM
0
100
200
300
400
500
Time (sec)
B
50pM CCK
n
s
c
ΔI/Io
=10%
0
100
200
300
400
500
600
Time (sec)

The $[Ca^{2+}]_i$ measurements were made using video imaging (QuantiCell, Applied Imaging, U.K.). To produce the result shown in Fig. 4B, the same agonist concentration was applied to a different cell, this time loaded with DTAF-CaM, with the result that we were able to resolve oscillations of calmodulin concentration with different patterns in different areas of the cell. Note how the $[Ca^{2+}]_i$ and calmodulin spikes in the SG region have a similar shape. In contrast, there was a sustained elevation of the nuclear calmodulin concentration with superimposed oscillations, which were delayed and incomplete compared to those in the SG region. This pattern is clearly different to the pattern of Ca^{2+} oscillations in the nucleus. The calmodulin concentration in the basal nonnuclear region also oscillated and appeared to be a mirror image of the oscillations seen in the SG region. Integration of Ca^{2+} spike-driven pulses of calmodulin entry by the nucleus could have important consequences for reactions driven by nuclear Ca^{2+}/CaM dependent enzymes, particularly Ca^{2+} dependent gene expression.

Troubleshooting

- Cells are washed away by the perfusion system or are lifted off the coverslip when removing the pipette:
  Prior to use, coat the coverslips that form the base of the cell bath with poly-L-lysine (Sigma) solution (diluted 1/10 with distilled water) for 10 minutes, rinse with distilled water and leave to dry. This should aid cell adhesion.

- How are estimations of the concentration of DTAF-CaM entering the cell made?
  The concentration of DTAF-CaM introduced to the cell can be estimated by comparing the fluorescence from the cell with the fluorescence recorded from a droplet of internal solution (K^+ HEPES) containing a calibrated amount of DTAF-CaM (i.e. 50 μM). In our experiments where 50 μM DTAF-CaM was present in the patch pipette, the concentration of DTAF-CaM in the pancreatic acinar cell was in the range of 20-30 μM. This is of the same order of magnitude as the estimated calmodulin concentration in the rat pancreas (Vandermeers et al., 1977).

Subprotocol 3
Monitoring the Ca^{2+}-Calmodulin Reaction in Different Subcellular Compartments

In order to observe the Ca^{2+} calmodulin reaction in different cellular regions we employed a different form of fluorescent calmodulin, TA-CaM, which increases its fluorescence upon calcium binding (Török and Trentham, 1994). The response of TA-CaM to cell stimulation with a calcium mobilising agonist has two components. Firstly, the fluorescence increase occurring after a rise in intracellular calcium concentration allows measurements to be made of the proportion of calmodulin that binds calcium, and is therefore activated, in each area of the cell. Secondly, TA-CaM redistribution occurs, and it is possible to calculate the amount of calmodulin that translocates to the nucleus and SG region, and the amount that leaves the basal nonnuclear area.

Procedure

The technique for introducing TA-CaM into the cell is the same as that already described for DTAF-CaM. However, as the TA-CaM response is more complex than DTAF-CaM, a few points will be made in this section concerning data analysis and calculations that may be carried out.

After loading the cell with TA-CaM (at a concentration of 100 µM in the patch pipette) remove the pipette, and start confocal imaging.

The recording conditions used for the experiment shown in Fig 5 were as follows: 100 confocal images were taken, exciting TA-CaM with the 364 nm wavelength produced by the UV laser, with a 400 nm barrier filter. The time required to scan one pixel was 400 nanoseconds (512 x 512), with 8.5 second intervals between each frame, using x32 averaging. After obtaining several images of the TA-CaM distribution in the cell at rest CCK (1 nM) was applied and fluorescence intensity was measured in the nucleus. (TA-CaM in an unstimulated cell has a resting fluorescence intensity higher than that of autofluorescence, and can increase its fluorescence approximately 5-fold after binding cal-

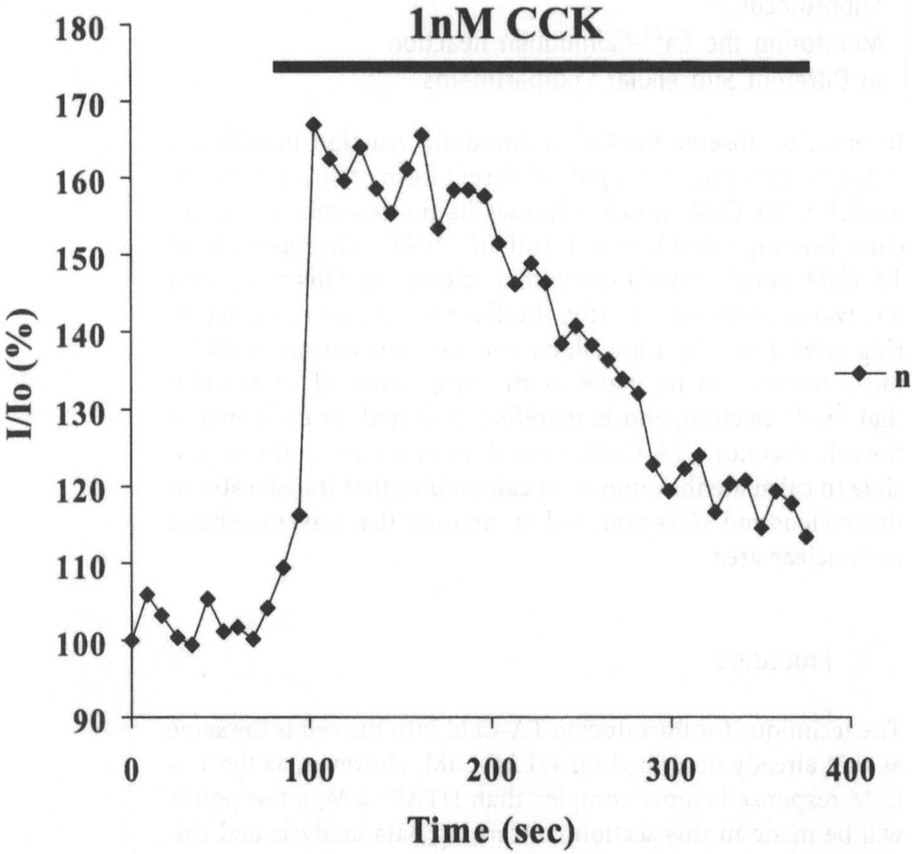

Fig. 5. A supramaximal dose of CCK induces rapid fluorescence changes in the nucleus of a TA-CaM loaded cell.

cium (Török and Trentham, 1994; Török et al., 1998) although in our hands, the changes were usually smaller than this).

Monitoring the calcium calmodulin reaction

It is possible to estimate the proportion of calmodulin that binds calcium as a result of agonist stimulation by comparing the TA-CaM fluorescence response to the agonist with the change of TA-CaM fluorescence induced by saturating the probe with calcium in a cell free system. In order to do this, it is necessary to find the minimum (Ca^{2+}-free form) and maximum (Ca^{2+} -bound) TA-CaM fluorescence by recording the intensity of small droplets of the probe in K^+ HEPES based solution containing ETGA (minimum fluorescence intensity) or Ca^{2+} (maximal fluorescence intensity).

1. The first step is to prepare the siliconized coverslips. Clean the surface of the coverslip by rinsing coverslips five times with distilled water, once with ethanol and then once again in water, directing the flow against the tilted coverslip so that droplets collect on the edge and can be gently wiped away.

2. Immerse coverslip in Sigmacote for 5 minutes.
   Sigmacote consists of a special silicone solution in heptane that forms a microscopically thin film on the glass. Since it repels water, it prevents droplets of solution placed on the glass from spreading, and is useful if the experimenter has to work with very small volumes of solution.

3. Dilute TA-CaM to a concentration of 50 µM in K^+ HEPES (pH 7.2) containing either 2 mM EGTA, or 1 mM Ca^{2+}, to give 20 µl of each solution.
   Using a small plastic pipette tip, deposit four small droplets of each solution in rows on the glass coverslip. Cover each row of droplets in a layer of mineral oil to prevent evaporation.
   The exact size of the TA-CaM droplets is not important when using a confocal microscope because a section will be taken through each droplet and fluorescence intensity in this section will not be influenced by out of focus light. This calibration could not be performed on a conventional fluorescence microscope with a single excitation wavelength indicator such as TA-CaM because the thickness, or depth, of individual droplets would affect the result.

4. Use the 364 nm laser line and recording conditions that are as close as possible to those used during a typical experiment with TA-CaM loaded cells to record the intensity of fluorescence from each droplet.
   This information is used to calculate the ratio of the maximal intensity of fluorescence of TA-CaM (Ca^{2+}-bound form) to the minimal intensity of fluorescence (Ca^{2+}-free form). For our experiments, this value, referred to as k, was 2.3. The following formula can then be used to calculate the proportion of calmodulin that binds calcium:

$$R = (\Delta F/F_0)/(k-1)$$

where R is the ratio of the concentration of TA-CaM that binds Ca^{2+} as a result of stimulation and the total concentration of TA-CaM in the cell. The concentration of TA-CaM in the cell can be estimated by comparing the fluorescence of a droplet of TA-CaM (at the same concentration used in the patch pipette) with the fluorescence measured in a TA-CaM loaded cell, using the same recording conditions, i.e. laser intensity and gain etc. The change in intensity of the whole cell fluorescence as a result of agonist application is represented by *F*, while F_0 denotes the initial intensity of fluorescence. We estimate that in our experiments, 24 ± 3% (n = 11) of TA-CaM binds Ca^{2+} as a result of agonist stimulation. If one corrects for the slight spectral shift (see troubleshooting), this value decreases to 23%.

Results

Supramaximal doses of CCK (or ACh) result in increased fluorescence of TA-CaM in all regions of the cell, but the nuclear response (Fig. 5) is larger in amplitude and fluorescence remains elevated for longer periods than the SG or basal non nuclear region response (data not shown). This suggests that the prolonged nuclear response may be due to accumulation of Ca^{2+}-bound calmodulin and it is possible to calculate the component of the fluorescent signal that occurs as a result of calmodulin redistribution.

Troubleshooting

- Preventing calcium contamination:
  TA-CaM is sensitive to calcium so calcium contamination occurring while TA-CaM is still in the pipette could affect the results of the experiment. Therefore, it is advisable that in addition to the 0.1 mM EGTA in the K^+ HEPES solution, the external medium used to perfuse the cells should be made without $CaCl_2$ and 50 μM EGTA should be included.

This will chelate the background level of calcium contamination found in any 'nominally calcium free' solution, and in the event of the giga seal becoming 'leaky', will prevent calcium binding to and activating the TA-CaM in the patch pipette.

- The maximal excitation wavelength for TA-CaM shifts from 380 to 365 nm on binding calcium, therefore the use of a single 364 nm excitation wavelength will overestimate Ca^{2+} binding.
  The spectra of TA-CaM (both Ca^{2+}-free and Ca^{2+}-bound) are quite broad, but there is a slight shift of the maximal excitation wavelength for TA-CaM upon binding Ca^{2+}. The effect of this spectral shift on the amplitude of the signal is however, quite small at an excitation wavelength of 364 nm. We estimate that at 364 nm the spectral shift could account for a 4.7% change in the amplitude of the fluorescence intensity. As a result the proportion of TA-CaM that binds Ca^{2+} could be overestimated by approximately 1%.

Acknowledgements. I would like to extend special thanks to Dr Teruko Takeo who worked closely with me for many of these experiments, to my supervisor Dr A.V. Tepikin, and to Professor O.H. Petersen for their support. We are extremely grateful to Dr K. Török for supplying the DTAF- and TA-CaM, and for the advice she was able to offer regarding their application. I would also like to thank the Wellcome Trust for financial support. The results of experiments shown in this chapter which involve DTAF-CaM have been published as: Craske M, Takeo T, Gerasimenko O, Vaillant C, Török K, Petersen OH & Tepikin AV, (1999) Hormone-induced secretory and nuclear translocation of calmodulin: Oscillations of calmodulin concentration with the nucleus as an integrator. Proceedings of the National Academy of Science U.S.A. 96 (8):4426-4431.

References

Bachs O, Agell N, Carafoli E (1994) Calmodulin and calmodulin-binding proteins in the nucleus. Cell Calcium 16:289-296

Bito H, Deisseroth K, Tsien RW (1997) Ca^{2+}-dependent regulation in neuronal gene expression. Curr Opin Neurobiol 7:419-429

Cheung WY (1970) Cyclic 3', 5'-nucleotide phosphodiesterase: Demonstration of an activator. Biochem Biophys Res Commun 33:533-538

Craske M, Takeo T, Gerasimenko O, Vaillant C, Török K, Petersen OH, Tepikin AV (1999) Hormone-induced secretory and nuclear translocation of calmodulin: Oscillations of calmodulin concentration with the nucleus as an integrator. Proc Natl Acad Sci USA 96 (8):4426-4431.

Dash PK, Karl KA, Colicos MA, Prynes R, Kandel ER (1991) cAMP response element binding protein is activated by Ca^{2+}/calmodulin – as well as cAMP-dependent protein kinase. Proc Natl Acad Sci U.S.A. 88:5061-5065

Deisseroth K, Heist EK, Tsien RW (1998) Translocation of calmodulin to the nucleus supports CREB phosphorylation in hippocampal neurons. Nature 392:198-202

Finkbeiner S, Greenberg ME (1997) Spatial features of calcium-regulated gene expression. BioEssays 19 (8):657-660

Harper JF, Cheung WY, Wallace RW, Huang HL, Levine SN, Steiner AL (1980) Location of calmodulin in rat tissues. Proc Natl Acad Sci USA 77(1):366-370

Kakiuchi S, Yamazaki R, Nakajima H (1970) Properties of a heat-stable phosphodiesterase activating factor isolated from brain extract: studies on cyclic 3', 5'-nucleotide phosphodiesterase. Proc Jpn Acad 46:587-592

Liao B, Paschal BM, Luby-Phelps K, (1999) Mechanism of Ca^{2+}-dependent nuclear accumulation of calmodulin. Proc Natl Acad Sci USA 96(11):6217-6222

Luby-Phelps K, Hori M, Phelps J, Won D, (1995) Ca^{2+} -regulated dynamic compartmentalization of calmodulin in living smooth muscle cells. J Biol Chem 270(37):21532-21538

Matthews RP, Guthrie CR, Wailes LM, Zhao X, Means AR, McKnight GS (1994) Calcium/calmodulin-dependent protein kinase types II and IV differentially regulate CREB-dependent gene expression. Mol Cell Biol 14(9):6107-6116

Mayford M, Wang J., Kandel ER, O'Dell TJ (1995) CamKII regulates the frequency-response function of hippocampal synapses for the production of both LTD and LTP. Cell 81:891-904

Petersen CCH, Toescu EC, Petersen OH (1991) Different patterns of receptor-activated cytoplasmic Ca^{2+} oscillations in single pancreatic acinar cells: dependence on receptor type, agonist concentration and intracellular Ca^{2+} buffering. EMBO J 10:527-533

Pruschy M, Ju Y, Spitz L, Carafoli E, Goldfarb DS (1994) Facilitated nuclear transport of calmodulin in tissue culture cells. J Cell Biol 127(6 part 1):1527-1536

Pujol MJ, Soriano M, Aligué R, Carafoli E, Bachs O, (1989) Effect of α-adrenergic blockers on calmodulin association with the nuclear matrix of rat liver cells during proliferative activation. J Biol Chem 264(32):18863-18865

Thorn P, Lawrie AM, Smith PM, Gallacher DV, Petersen OH (1993) Local and global cytosolic Ca^{2+} oscillations in exocrine cells evoked by agonists and inositol trisphosphate. Cell 74:661-668

Török K, Trentham DR (1994) Mechanism of 2-Chloro-(ε-amino-Lys_{75})-[6-[4-(*N*,*N*-diethylamino)phenyl]-1,3,5-triazin-4-yl]calmodulin interactions with smooth muscle myosin light chain kinase and derived peptides. Biochemistry 33:12807-12820

Török K, Wilding M, Groigno L, Patel R, Whitaker M (1998) Imaging the spatial dynamics of calmodulin activation during mitosis. Current Biol 8:692-699

Vandermeers A, Vandermeers-Piret MC, Rathe J, Kutzner R, Deforge A, Christophe JA (1977) A calcium dependent protein activator of guanosine 3', 5'-monophosphate phosphodiesterase in bovine and rat pancreas. Eur J Biochem 81:379-386

Suppliers

Katalin Török (DTAF & TA-CaM)
School of Biological Sciences
Queen Mary & Westfield College
University of London
London, E1 4NS
k.torok@qmw.ac.uk

Acrodisc (filters)
Gelman Sciences,
Michigan, U.S.A.

Sigma-Aldrich Company Ltd. (general reagents for solutions, etc.)
Fancy Road
Poole, Dorset, BH12 4QH
England
Phone: 01202 733114

Molecular Probes (hoechst, ionomycin)
4849 Pitchford Avenue
Eugene
Phone: (541) 465-8300
Fax: (541) 344-6504
Europe: PoortGebouw,
Rijnsburgerweg 10
2333 AA Leiden
The Netherlands
Phone: +31-71-5233378
Fax: +31-71-5233419
http://www.probes.com

Worthington Biochemical Corporation (collagenase)
730 Vassar Ave,
Lakewood, NJ 08701
Phone: 800-445-9603
Fax: 800-368-3108

Abbreviations

CaM	calmodulin
SG	secretory granules
$[Ca^{2+}]_i$	cytosolic calcium concentration
CCK	cholecystokinin
ACh	acetylcholine
TA-CaM	2-chloro-(ε-amino-Lys_{75})-(6-(4-N,N-diethylamino-phenyl)-1,3,5-triazin-4-yl)-CaM
DTAF-CaM	5-(4,6-dichlorotriazinyl)-amino-fluorescein-CaM
EGTA	ethylene glycol-bis(β-aminoethyl ether)-N,N,N',NN'-tetra-acetic acid

Subject Index

A
acetoxymethylester 10
acetylcholine 211
acinar cells, pancreatic 212, 224, 249, 263, 268, 269
acoustooptic device 18
acquisition speed 43
ADC (analog-to-digital converter) 34, 38
adenylate cyclase 265
adrenal chromaffin cells 187
ADU (analog-to-digital unit) 35
aequorin 65–86, 105, 107, 198, 201
–, antibodies raised against 122
–, chimeric 112, 115, 146, 201, 204
– – cytosolic 116
– – nuclear 116
–, Ca^{2+}-affinity 147, 154
–, co-expressed 111
–, consumption 159
–, dialyisis buffer (ADB) 70
–, ER-targed 145, 147
–, functional 119
–, luminescence
– – Ca^{2+}-dependence 154
– – calibration data 82
–, measurements/measuring systems 120
– – cytosalic Ca^{2+} 65–86
–, mitochondrial calcium measurements 198, 201
–, mutated 152, 156
–, preparaiton of solution for dissolving 70
–, recombinant 111, 123
–, reconstitution 106, 146
–, specific localisation 108
agarose gel 65
air isolation table 272
α-amylase 217, 218
aluminosilicate 93
analog-to-digital
–, converter (ADC) 34, 38
–, unit (ADU) 35
anode current 24
antibodies raised against aequorin 122
apoaequorin 107, 146
apoprotein, recombinant 119
apoptosis 131
area analysis 57
argon
–, ion 18
–, krypton 18
–, laser 7, 42
astrocytes 187
ATPase
–, Ca^{2+} ATPase 249, 265
–, ER Ca^{2+} ATPase 217, 224
–, proton ATPase 229, 230, 237, 239
– – vacuolar-type 229, 237, 243
atropine 125

B
backround fluorescence 13
bafylomycin 229–231, 237, 243
bandpass filter 15
bandwidth 15, 17, 18
BAPTA (*see also* oregon-green) 188
–, BAPTA-1 5
–, BAPTA-5N 5, 167, 234–236, 238
BCECF 19

beamsplitter 25
BHQ 148
binning 28, 36, 37
–, pixel binning mode 36
bis-fura-2 5
black level 35
bleaching 233
blooming 29
*.bmp (bit-mapped images) 47, 51
borosilicate
–, glass
– – capillaries 74
– – tubing 93
–, micropipettes 68, 71, 74
bradykinin 174
brain slice preparations 11
buffer/buffering effect 12, 109, 110, 170
–, high molecular weight 258
–, mobile calcium buffer 265

C
Ca^{2+} (calcium)
–, activity 91
–, aequorin affintiy (*see there*) 147, 154
–, AT-phases 110
–, ATPase (*see there*) 249, 265
–, buffering (*see there*) 12, 109, 110, 170
–, caged (NP-EGTA) 241, 242, 259, 261
– – flash photolysis 241
–, calcium-*Crimson* 5
–, calcium-green 5, 7, 133, 135
– – 1-AM 134
– – calcium-green-1 5
– – calcium-green-1-dextran 136, 222, 223, 258–260
– – calcium-green-2 5
– – calcium-green-5N 5, 8, 257
– – Mag-Fura2-AM 134
–, calcium-orange 5, 7
– – calcium-orange-5N 5, 8
–, calibrations 230
–, calmodulin (*see* there) 265–288
–, channels 110
–, Cl^- or K^+ currents, Ca^{2+}-sensitive 173
–, cocktail, Ca^{2+}-ligand 93
–, contamination 284, 285
–, cytosalic, aequorin measurements 65–86
–, dextran-bound calcium indicators 142
–, dye (*see* dyes)
–, efflux 249
–, extrusion (*see there*) 244, 249, 258
–, indicators
– – non-ratiometric 5
– – high-affinity 5
– – low-affinity 5
– – ratiometric 5
– – high-affinity 5
– – low-affinity 5
–, intragranular free Ca^{2+} concentration 220
–, ion selective electrode 257
–, measurement/measuring
– – in endosomes of intacts cells 229–244
– – extrusion from single cells 249–263
– – inside organelles 103–127
– – of internal store 172
– – with fluorescent probes in the endoplasmic reticulum 163–180
– – in isolated nuclei 131–143
– – with luminescent probes in the endoplasmic reticulum 145–161
– – in mitochondria 185–207
– – in secretory granules 211–224
– – synchronous measurement of intracelluler and extracellular calcium concentrations 250, 255
–, microdomains 204
–, microelectrodes, Ca^{2+}-sensitive 91
–, nuclear Ca^{2+} transport, model 143
–, nucleoplasm of intact cells, Ca^{2+}-measuring 105–127
–, oscillation 249

–, phosphate
– – co-precipitation 118
– – procedure 117, 118
–, pump 131, 244
–, secretion, rate of 258
–, sponge S 170
–, uniporter 205
cADP-r (cyclic adenosine diphosphate-ribose) 133, 140, 142, 158, 211, 216, 219
caffeine 97, 157, 158
caged calcium in endosomes, flash photolysis 241
calcein 19
calcium (*see* Ca^{2+})
calculation, Ca^{2+} concentration 255
calibration
–, of Ca^{2+} concentration 230, 255
–, data, aequorin luminescence 82
–, dyes 12
–, luminescent data 151
–, pH 230
–, ratiometric probe mag-fura-2, calibration procedure for 178
–, raw luminescent signal 121
–, rhod-2 193
–, in situ 176
–, solutions 94
–, in vitro 176
calmodulin (CaM) 131, 163, 265
–, binding reaction 266
–, Ca^{2+}-dependent 265
–, fluorescently labelled 266
–, injection
– – of DTAF-CaM 269, 270
– – of labelled calmodulin 268
–, kinase 266
–, monitoring Ca^{2+}-calmodulin reaction 281, 282
–, movements 273
–, nuclear CaM concentration 280
–, oscillations of CaM concentration 278
–, redistribution 275
–, solution 273
–, TA-CaM 281–285
–, tracking inside cells 265–288
–, translocation of 265
calreticulin 131
cameleons, ER-targeted 175
camera(s) 25–40
–, camera-based imaging 21
–, CCD 4, 39
–, transfer 27
– – B2 interline transfer 27, 28
– – frame transfer 27
– – full frame transfer 27
carbachol 125, 202
cardiac myocytes 66, 187
CCD (charged-couple device)
–, array 38
–, camera 4, 39
–, chips 22, 25, 26, 34, 36, 39, 47
–, detector 32
–, interline CCD detectors 28
–, sensor 39, 40
–, temperature 29
cDNA
–, coding regions 116
–, non-coding regions 116
cell(s)/cellular
–, adrenal chromaffin cells 187
–, attachment 160
–, cultures 117
– – single cells 8
–, cycle regulation 131
–, estimation of volumes 256
–, HeLa cells 122, 124, 152, 202, 204
–, loading 10
–, pancreatic acinar cells 212, 249, 263, 268, 269
–, patch clamp technique/configuration, whole cell 173, 268, 270
–, permeabilization 169, 180
–, rapid changes of the cell bath fluid 267
–, skeletal muscle cells 118, 125
–, subcellular fractionation 201
–, suspensions 8
–, swelling 273
charged-couple device (*see* CCD)
chelating column 170
cholecystokinin (CCK) 211
chopper mirrow 18
chromaffin cells, adrenal 187
citrate synthase 201

Cl^- or K^+
–, currents, Ca^{2+}-sensitive 173
–, release 243
clamp the cytoplasmic Ca^{2+} 11
clones, stable 118
cocktail, Ca^{2+}-ligand 93
coelenterazine 105, 107, 108, 119, 120, 146, 148, 150, 154
compartmentalisation/compartmentalized 10
–, low-affinity probes 164
compression 50
confocal microscope/microscopy 4, 41–43, 63
–, conventional 41
–, laser scanning microscope (CLSM)/laser systems 21, 22, 42
–, multi-photon 43
–, two-photon 41
contrast enhancement 55, 61
Corel drawing 47
cortical neurons, primary cultures 118, 126
cover-glasses, preparation of 251
CREB 266
culture/cultured
–, cells 8, 117
– – single cells 8
–, conditions 119
–, primary 118
– – cortical neurons 118
– – skeletal muscle 118, 125
cyclic
–, adenosine diphosphate-ribose (cADP-r) 133, 140, 142, 158, 211, 216, 219
–, nucleotide phosphodiesterase 265
cytochrome c oxidase 199
cytoplasmic Ca^{2+}
–, clamp the 11
–, extrusion of cytoplasmic dye 174
cytosolic
–, aequorin chimeras 116
–, calcium spikes 211

D
3D reconstructions 23
dark current 29–31
–, current 31
–, signal 31
detection
–, sensitivity 22
–, system 14
detector 4
dexamethasone 123–126
dextran 132
–, calcium indicators, dextran-bound 142
–, fura-2–dextran 135, 136, 137, 142
dialysis
–, apparatus 72, 73
– – microdialyis 72
–, intracellular 12
dichroic(s) 25
–, emission 19
–, filter dichroic/emission combination 19
–, mirrow 13, 14, 19
–, turret 20
DICOM (digital imaging and communications in medicine) 54
diffraction grating 18
digitonin 122, 159, 171, 205
dihydroRhod-2 AM 189, 192
$DiOC_6(3)$ (ER [endoplasmic reticulum] marker) 132, 136, 142, 217
DNA
–, binding 116
–, repair 131
–, replication 131
–, stain 274
droplet
–, estimation of volumes 256, 257
–, experiment 250
–, formation 251, 252
–, measurements 254
–, technique 249, 250
DTAF-CaM 269–280
–, injection 269, 270
–, intracellular distribution 273
dyes
–, affinity of 6

–, calibration 12
–, cytoplasmic, extrusion of 174
–, dual-emission 25
–, dual-excitation 6
–, non-lineary dye response 176
–, non-ratiometric 7
–, quenching of dye fluorescence 179
–, ratiometric 5–7, 18, 19, 62
–, single wavelength 18, 165
–, spectra of 165
dynamic range (DR) 29, 30, 35, 37, 109

E
EF-hand 106, 108
EGTA 150, 151, 159, 160
electrical
–, interference 92
–, resistance 91
electroneutral antiporter 185
electrophoretic uniporter 185
electroporation 112, 118
electrostatic interference 100
emission
–, dichroic 19
–, dual 19, 25
–, low-light 111
–, multiple 20
–, single 19
–, spectrum 6, 17
-rhod-2 190
encapsulated PostScript (*.eps) 51
endocytosis 231, 232
–, labelling, endocytic 232
–, rate of 240, 241
endoplasmic reticulum (*see* ER)
endosomes/endosomal
–, acidification 237, 243
–, Ca^{2+}
– – caged Ca^{2+}, flash photolysis 241
– – concentration 236, 237
– – release 234, 240
–, early endosomes 229
–, free Ca^{2+} concentration 232
–, of intacts cells, Ca^{2+} measurements 229–244
–, lumen of 229
–, pH 232, 235
–, proton accumulation 243
epitope tag 122
*.eps (encapsulated PostScript) 51
ER (endoplasmic reticulum) 212, 217
–, aequorin, ER-targeted 145
– – chimeric mutated (ERmutAEQ) 146, 147, 149, 159
–, Ca^{2+} 165
– – ER Ca^{2+} ATPase 217, 224
–, cameleons, ER-targeted 175
–, fluorescent probes, Ca^{2+}-measurement 163–180
–, marker $DiOC_6(3)$ 136, 137, 142, 217
esterases 10
ethidium bromide 135
excitation
–, dual 19
–, energy 17
–, filter 4, 254
– – interference 16
–, light 19
–, multiple 19, 20
–, spectra/spectrum 166
– – separation of 255
–, wavelengths 6
exocytosis 229
–, agonist-evoked 244
–, calcium efflux by 249
extragranular medium 218
extrusion from single cells, Ca^{2+} measuring 249–263
–, Ca^{2+} dependency of Ca^{2+} extrusion 258
–, visualization of calcium extrusion 249
eyepiece 14

F
Faraday cage 273
F_i/F_0 ratio images 62
fibroblast 234, 236, 238
file
–, formats 49–54
– – general 49
– – proprietary image formats 49
–, header 49

filters
–, bandpass 15
–, dichroic/emission combination 19
–, edge-detection 60
–, excitation (*see there*) 4, 16, 254
–, infrared blocking (IR) 15, 16
–, longpass 15, 19
–, multiband 15, 20
–, neutral density (ND) 15, 17
–, sharpening 59
–, smoothing 60
–, *Sobel* 60
–, spatial 58
–, wheel 16, 19, 20
– – rotating 16
FITC-dextran 233
fluo-3 5, 7, 8, 215, 255
–, fluo-3 – ff 165, 167
fluo-4 7
fluorescein/fluorescent 19, 272
–, calmodulin, fluorescently labelled 266
–, green fluorescent proteins 163
–, low-affinity indicators 163, 164, 167
–, lysotracker fluorescence 217
–, probes in the endoplasmic reticulum, Ca^{2+}-measurement 163 – 180
–, quenching of dye fluorescence 179
–, targeted fluorescent indicators 163
free Ca^{2+} concentration 220
–, in endosomes 232
–, intragranular 220
fura-2 3, 5, 7, 8, 19
–, fura-2-AM 132
–, fura-2-dextran 135, 136, 137, 142
–, fura-2-ff 165, 167
–, mag-fura-2 5, 8, 165, 167, 178, 257
–, mag-fura-5 5, 8, 165, 167
–, mag-fura-2-AM 134, 174
–, mag-fura-red 165, 221
–, mag-fura-red-AM 214, 215
fura red 5, 7

G
gain 31, 35
γ correction 35
ganglion, isolated 99
gap junctions 272
gene
–, expression 131, 265
– – regulation of 265
–, gun 112
*.gif (graphics interchange format) 52
giga seal 274
glucocorticoid receptor (GR) 115, 116
glutamate 126, 202
granules
–, intragranular free Ca^{2+} concentration 220
–, isolation of 213
–, sectretory granules, Ca^{2+} measurement (*see there*) 211 – 224
–, single granules 222
–, zymogen (ZGs) 212, 213, 216, 219, 221, 221
gun
–, gene gun 112
–, particle gun delivery 118

H
H^+ accumulation 230
heat-shock protein 115
heavy metals 179
–, chelator 179
HeLa cells 122, 124, 152, 202, 204
helium-neon 18
heparin 219
hepatocyte 84, 85, 187
high
–, dynamic range mode 36
–, sensitivity mode 36
–, speed sampling 16
hippocampal neurons 266
histamine 123, 202, 204

I
image
–, analysis 54
–, background image (B) 55
– – B corrections 56

–, *.bmp (bit-mapped images) 47, 51
–, capture stage 54, 55
–, digital imaging and communications in medicine (DICOM) 54
–, display 54
–, experimental (E) 54
–, F_i/F_0 ratio 62
–, file formats (*see there*) 49-54
–, intensifier 25, 38
–, mask 61
–, plane 4, 13
–, processing 55, 56
– – basic image processing 54-64
–, processor 4
–, scan 23
–, segmentation 56, 60
–, *.tif (tagged-image file format) 50, 53
–, vector-based 47
immunocytochemical localisation 123
immunofluorescence data 15, 200, 205
immunolocalization 200-202
immunological tag 117
incident light 15
indo-1 5, 6, 8, 19, 25, 215
infrared
–, blocking (IR) filter 15, 16
–, laser, pulsed 43
injection
–, DTAF-CaM injection 269, 270
–, IP_3 injection 212
–, labelled calmodulin 268
–, microinjection (*see there*) 65-68, 70, 74, 75
–, pressure injection 253
inositol 1,4,5-triphosphate (IP_3) 131, 142, 211, 216, 219
$InsP_3$ 220, 223
integration time 29
interlaced picture 26
interline transfer camera 27, 28
internal store [Ca^{2+}] measurement 172
intragranular free Ca^{2+} concentration 220
ionomycin 138, 168, 215, 231, 235, 238
ionophores 176, 235
IP_3 (inositol 1,4,5-triphosphate)
isolation of the granules 213

J
JC-1 19
*jpg (joint photographic expert group) 53

K
K^+ or Cl^-
–, currents, Ca^{2+}-sensitive 173
–, release 243

L
laser
–, argon-laser 7, 42
–, confocal systems 21, 22
–, pulsed infrared 43
–, UV-laser 18, 43
LDH (lactate dehydrogenase) 201
light
–, damage 233
–, excitation 19
–, guides 25
–, incident 15
–, low-light emission 111
–, source 4
–, transmitted light picture 216
–, UV-light source 25
liposomes 112
liver, isolated nuclei 132, 137
longpass filter 15, 19
low-affinity probes 163, 164
–, compartmentalized 164
–, fluorescent indicators 167
lumincescence/luminescent 106
–, luminescent probes in the endoplasmic reticulum, Ca^{2+} measurement 145-161
–, protein 65
–, raw signal, calibration 121
luminometer 120
LUT (lookup table) 48, 61, 62
lymphocytes 187
lysotracker fluorescence 217

M
mag-fura-2 5, 8, 165, 167, 257
–, mag-fura-2-AM 134, 174
–, ratiometric probe mag-fura-2, calibration procedure for 178
mag-fura-5 5, 8, 165, 167
mag-fura-red 165, 167, 221
–, mag-fura-red-AM 214, 215
mag-indo-1 5, 8, 165, 167
marine eggs 131
mask (M) 55
MaxChelator 169
measurement artefact 132
meiosis 131
mercury 16
metallochromic indicators 105
Mg^{2+} sensitivity 166, 180
microdialyis apparatus 72
microelectrode
–, Ca^{2+}-ion selective electrode 257
–, Ca^{2+}-sensitive 91
– – construction 95
–, glass 93
–, for microinjection 67
–, puller 70, 74, 94
– – vertical 70, 74
microinjection 65
–, apparatus 66, 75
–, microelectrode for (*see there*) 67, 70, 74
–, pipettes for 68
micropipette
–, borosilicate 68, 71, 74
–, filling 94
–, IP_3 injection 212
–, silanise 93, 95, 96
microplate channel 39
microscope/microscopy/microscopic
–, confocal (*see there*) 4, 18, 41–43, 63
–, epifluorescent 13
–, inverted 13
–, real-time confocal 40–44
–, upright 13
microslides 65–70, 78, 80
microsomal suspensions 8
mirrow
–, chopper 18
–, dichroic 13, 14
–, rotating 42
–, scanning 42
misellaneous 247–288
mitochondria 109
–, aequorin 201
–, homeostasis 186
–, intermembrane space 199
–, matrix 197, 199
–, measurements 185–207
– – using aequorin 198
–, responses 202
–, rhod-2 loading into mitochondria 190
–, uptake 185
mitosis 131
mobile messenger 212
monitoring Ca^{2+}-calmodulin reaction 281, 282
monochromator 4, 18–20
–, tuneable 16
multiband filter 15
muscle cells
–, skeletal muscle cells 125
–, smooth muscle cells (*see there*) 188
mycocytes 202
myotubes 125

N
Na^{+}HEPES 267, 268
NAADP 211
Nernst equation 91
neurons 187, 202, 249
–, cortical neurons, primary cultures 118, 126
–, hippocampal 266
neutral density (ND) filter 15, 17
nigericine 231, 235, 238
Nipkow disk 41
noise 29
–, dark noise
– – current 31
– – signal 31
–, photon-shot 29
–, read-noise limited 29
–, readout 29, 31
–, signal-to-noise ratio (S/N) 29, 30, 35, 38, 110
non-interlaced picture 26

non-ratiometric Ca^{2+}
–, dyes 7
–, indicators (*see* Ca^{2+}) 5
NP-EGTA (caged-calcium) 241, 259, 261
–, flash photolysis 242
nuclear
–, aequorin chimeras 116
–, Ca^{2+} transport, model 143
–, CaM concentration 280
–, envelope 113, 121, 131, 132, 139-142
- breakdown of 121
–, fundamental nuclear processes 131
–, localisation sequences 114, 117, 132
–, membranes 114
- inner 114
- outer 114
–, perinuclear space 114
–, pore complex 114, 131
–, targeting 113
–, translocation 115, 125
nucleoplasm 121, 132, 142
–, nucleoplasmic space 141
nucleoporin 132
nucleus/nuclei 212, 265
–, isolated nuclei, Ca^{2+} measuring 131-143
- liver 132, 137
- single nuclei 138, 141
–, location 274
–, translocation in 115
numerical aperture (NA) 31, 32

O
oligodendrocytes 187, 188
optical efficiency 18
oregon-green 5, 167
–, 488 BAPTA-1 5
–, 488 BAPTA-5N 5, 167, 234-236, 238
oregon-orange-5N 5
oscillation 249
–, of CaM concentration 278
oven
–, silanisation 94
–, vacuum 94

P
pancreas 216
–, acinar cells, pancreatic 212, 224, 249, 263, 268, 269
particle gun delivery 118
patch
–, clamp technique/configuration, whole cell 173, 268, 270
–, pipette 173, 196, 272, 273
- resistance 273
perfusion
–, chamber 78
–, intracellular 10
perinuclear space 114
permeabilizing
–, agents 168
–, cells 169
–, process 171
pH
–, calibration 230
–, endosomal 232, 235
–, measurements 19
–, probe, PH-sensitive 234
phosphatase 265
phosphor
–, channel 39
–, screen 39
photobleaching 275
photocathode 23
photodiodes, silicon 25
photomultiplication process 23
photomultiplier 120, 149
–, 9789A 149
–, housing 79
–, tube (*see* PMTs) 4, 21-25, 79
photon
–, limited 29
–, multiple-photon systems 41
photon-shot noise 29, 30
photoprotein 109
–, affinity 109
–, consumption 108
–, retention 146
–, targeting 146
pin-hole 18, 42
pipette
–, for microinjection (*see* micropipette) 68, 71, 74, 93-96, 212
–, patch pipette 173, 196, 272, 273

– – resistance 273
–, plastic pipettes 251
pixel 28–31
–, binning mode 36
–, depth 48
–, mean pixel intensity 57
–, size 28, 33
–, superpixel 37, 38
–, well capacity 28, 30
plasma membrane Ca^{2+} pumps 244
plastic pipettes 251
X-t plot 63
pluronic F-127 170
PMTs (photomultiplier tube) 4, 21–25, 42, 79
–, measuring system 25
–, spectral response 24
pore-forming toxin 168
pressure injection 253
protein
–, apoprotein (*see there*) 119
–, content 218
–, green fluorescent proteins 163
–, heat-shock protein 115
–, luminescent 65
–, photoprotein (*see there*) 109, 146
–, total protein concentrations 217
proton
–, accumulation 230, 243
–, ATPase (*see there*) 229, 230, 237, 239
–, gradient 205, 206
pseudo-coloured 48
pulmonary artery smooth muscle cell 195
pulse(photon)-counting 24
pulsed infrared laser 43

Q
quantum efficiency (QE) 28, 31
quartz tubing 101
quenching of dye fluorescence 179
quin-1 4

R
ratiometric Ca^{2+}
–, calibration procedure for ratiometric probe mag-fura-2 178
–, dyes 5–7, 18, 19, 62
–, indicators (*see* Ca^{2+}) 5
–, wavelength indicators 165
ratioing 56
readout
–, modes 27, 30
–, noise 31
–, rate 33
–, speed 31
real-time confocal microscopy 40–44
recombinant
–, aequorin 111, 123
–, apoprotein 119
resolution
–, spatial 21, 31
–, temporal 22, 33, 35
rhod-2 5, 186–192, 196, 197, 257
–, calibration 193
–, dihydroRhod-2 AM 189, 192
–, emission spectra for rhod-2 190
–, loading into mitochondria 190
–, rhod-2 AM 189, 191, 197
rhod-5N 166
rhodamine 188
–, rhodamine 123 19
–, rhodamine B isothiocyanate dextran 238, 240
R_{max} 6, 12, 176
R_{min} 6, 12, 176
RNA 132
ruthenium red. 205
ryanodine receptors 142, 157
–, actions of 142

S
S/N-(signal-to-noise)-ratio 29, 30, 35, 38, 110
scan/scanning
–, image scan 23
–, line scan 23, 57
–, mirrows 42
–, modes 26
–, rate 33

secretory
–, granules, Ca^{2+} measurement 211–224
– – loading of 214
–, pole 212
selective targeting 112, 113
sensitivity range 109
SERCA inhibitor 173
shading and backround corrections 56, 57
signal-to-noise ratio (S/N) 29, 30
silanizing/silanisation 94, 99
–, micropipettes 93, 95, 96
–, oven 94
silicon/siliconization 253
–, coverslips, siliconized 283
–, intensified target (SIT) 25
–, photodiodes 25
single wavelength indicators 165
in situ calibration 176
skeletal muscle
–, cells 125
–, primary cultures 118, 125
slit opening 42
smooth muscle cells 188
–, pulmonary artery 195
snail neurone 97, 98
Sobel filter 60
spatial
–, convolution 59
–, filters 58
–, resolution 21, 31
spectrofluorimetry 22
spectrum
–, emission 6, 17
–, lines 18
Sr^{2+} 110
steroid
–, binding 116
–, hormone 116
streptolysin O 168
subcellular fractionation 201
superpixel 37, 38
synchronous measurement of intracelluler and extracellular calcium concentrations 250, 255

T
TA-CaM 281–285
thapsigargin 173, 220, 224, 244
*.tif (tagged-image file format) 50, 53
total protein concentrations 217
transcription
–, activation 121
–, factor 266
transfection techniques 112
transfer cameras 27
transient transfection 118
transmitted light picture 216

U
UV-
–, Ca^{2+} indicators 8
–, excitation experiments 16
–, laser 18, 43
–, light source 25

V
vacuolar-type proton ATPase 229, 237, 243
vector-based images 47
verapamil 240
vibration problems 20
video tube 25
viral infection 118
in vitro calibration 176

W
wavelengths
–, excitation 6
–, indicators 165
–, single wavelength dyes 18
–, tuning 18
well capacity 28, 30

X
xenon lamp 18
–, arc lamps 16
X-rhod-5N 166
X-t plot 63

Z
zink 179
zymogen granules (ZGs) 212, 213, 216, 217, 219, 221
–, isolated 221

MIX
Papier aus verantwortungsvollen Quellen
Paper from responsible sources
FSC® C105338
www.fsc.org

If you have any concerns about our products,
you can contact us on
ProductSafety@springernature.com

In case Publisher is established outside the EU,
the EU authorized representative is:
Springer Nature Customer Service Center GmbH
Europaplatz 3, 69115 Heidelberg, Germany

Printed by Libri Plureos GmbH
in Hamburg, Germany